Trusting the Journey

Tess Corps

For You,

Thank you for listening.

CONTENTS

ACKNOWLEDGMENTS

I thank from my soul the many pilgrims who have traveled The Way before me, I felt you all.

Carlota and Jorge, my forever Camino family.

Sarah Bruce, my daughter, for telling me to keep my stories for my book, so I would feel them.

Nancy Andersen, my first reader, and constant support.

Alan Des Harnais, fellow author and patient friend.

And

Him, for fucking up.

PROLOGUE

Dear Tess,

I know you are in a bad spot right now in your life. The questions running through your head are overwhelming your every thought. You had so much invested in the life you were living with Him and believed it was forever *until* He did the unthinkable. You have tried with all you are to rationalize His choices but have realized it was beyond your control. I know you are feeling lost and so very alone.

I am going to tell you a story that will change your life. A story about you, from me.

Sincerely,

Tess

I was surprised that day in February when he reacted to the movie we had just finished watching, *The Way*. He rarely had any kind of emotional attachment to anything, so when He did, I listened close.

"I can see you doing that," He said.

It was the beginning of a thought that would take me on a journey so very far away from anything I knew.

IT BEGAN

Three months later, tucked away in a friend's unfinished basement, I bought the plane ticket and trusted everything else would fall into place. Come September, I would board a flight from Vancouver, with only a purple backpack, and begin a pilgrimage from St-Jean-Pied, France to Santiago, Spain, an 830 kilometer walk, which dates back to Roman times.

I was drawn to the idea when, shortly after watching the movie *The Way*, He chose to dissolve what love I had left for Him. I could no longer roll along in the world I was desperate to exist in, I had lost who I was. The call to walk the Camino enthralled my desire for something more than I was living, but, more importantly, the challenge of spending six weeks alone with my own thoughts offered a ritualistic rite of passage to a better self. At 52 years of age, I wanted to take a good hard look at me and see where the second half of my life could take me. Sometimes drastic upheaval requires drastic moves, and walking across Spain was what I needed.

So I began to plan: first sharing with anyone who would listen to get their reactions that would push me forward into the unknown then liquidating what little I had left, throwing myself into hours of work as a waitress, and coming home crying my broken heart out. I started plowing through the slow recovery of my shattered life and to do this, I knew I needed to leap outside the comfortable box I allowed myself to live in for far too long. I was going to dare an experience of the unknown to alter my thoughts and hopefully put life into perspective, an amazing thing that happens when you begin the process of being honest with yourself.

I did not know if this pilgrimage was the answer, I just knew it was big enough to rattle me. I searched for years through self-discovery books, talked to others, took short road trips alone, tried to meditate, and drank

2

wine (lots of wine), and, from this, I would have moments of enlightenment but nothing would stick. Yet here I was embarking on another way of seeking happiness, it was all I could do. Maybe this time it would stick.

With each day bringing me closer to leaving, I was focused on what needed to be done first and what I would do when I got back—it became very surreal. Ahead of me was six weeks of my life on the other side of the world, all by myself walking and walking and walking...what the hell was I thinking? I am not a lazy person but I am the type to always find the closest parking spot. Thoughts would roam my head, telling me I should be in training for what lay ahead, but, then again, when had anything in my life trained me for what I had so far encountered. I did manage a walk around the block with my fully-loaded backpack, but only once. When I drove anywhere I would calculate the distance: mmm that's what five kilometers looks like, and if I walked to work that would be 15 kilometers—almost a day on the Camino! Alright then, I got it covered I figured as I calculated while sitting in my car with a smoke and a Timmies coffffee. That was my physical training.

Mentally, I journaled my thoughts every day, prepared music on my iPhone, did Yoga with my daughter, watched inspirational movies, and talked often about what I was going to do. The more I talked about the Camino the more I became convinced I knew what I was doing and where I was going. I read a ton of "What to bring on the Camino" lists, YouTubed advice of what to expect, and even attempted to read Shirley MacLaine's experience of the Camino, even though it just wigged me out. In the end, I bought proper hiking boots, a backpack, which I returned as it didn't feel right, another backpack, also returned because it was too small, a third backpack, returned because it was too big, and then, finally, back to the first backpack because it was purple and what the hell did I know?

With my meager funds, I also purchased expensive merino wool socks, as recommended, and a fancy lightweight sleeping bag. For my hiking attire, I forged through thrift stores and Walmart to find leggings, yoga pants, a skirt, a light raincoat, and a wick away t-shirt for two bucks, what a find! For toiletries—I never knew how much a bottle of conditioner weighed until I had to carry it. You are only supposed to carry 12 percent of your body weight, which meant I was only allowed 13 kilograms—needless to say, the toiletries became bare essentials: toothbrush and paste, face/body cream, shampoo, (conditioner was no longer an option), deodorant, hairbrush, and elastics. Once I accumulated all my supplies, I packed and repacked my bag many times. Should I roll? Should I fold? Should I separate into baggies, two panties or three, how much does that weigh?

How many fucking band-aids will I need? On the final pack, 10 hours before I left, I just stuffed all my shit in and weighed in at 13.2 kilograms, yes down to an exact science! The heaviest item was the journal that took me ages to pick out as I sought out the lightest one possible. Little did I know then that it would be the first of four I would carry the whole way.

September 26th, 2016 arrived. My supportive daughter, Sarah, took me to the airport in Vancouver. This was seriously happening. There was a ticket waiting for me to Paris, France, and all I could think was "Holy Crap, I am really doing this." Sarah and I hugged for the last time and off I went into the endless serpentine of the airport security line. Little ol' me in my Walmart yoga pants, two dollar t-shirt, hiking boots, and purple backpack. For every loop I took, I could still look through the doors and see my biggest fan, Sarah, waving and smiling until it was no longer possible. When she disappeared from sight, I felt alone but independent and stood taller than I had in a long time. It would be okay, I was simply going for a long walk.

Once settled in my seat on the plane, before I fell into a wine sleeping/pill-induced rest, I embraced my thoughts of positivity knowing with each day I would gravitate towards what my life could possibly be. I was digging deep to be excited about my future, hoping my heart would heal and that I would feel the joy of who I was. After three months of planning and thinking about the journey ahead, all that was left was to allow it to unfold. I was letting it go. I was trusting the journey. Why else would I be sitting on a plane to France, headed to the Camino, center aisle next to a farting man…

PARIS

After a ten hour flight, I landed in Paris groggy and hungry. Heading out of customs I found all I needed: a large coffee to invigorated me for the next phase of my adventure. I had six hours before the next flight and there was nothing I wanted more than to see the famous Eiffel Tower. The only hitch in the plan was that, within those six hours, I needed to be on the other side of Paris at a smaller airport to catch the next plane. Minor detail. French public transport, here I come!

I quickly learned I had zero comprehension of the true French language. I thought some high school French would kick in, but no way. I laughed at myself for assuming in a foreign country that everyone would understand me. Approaching each bus driver in the chaos of the bus corral area, asking "Eiffel Tower, s'il vous plait," I was rewarded with rapid never heard before French. The gist of each response translated something like this, "Oh God another Canadian that thinks we speak English and we don't and the bus you want is…" I was not sure how I finally found the correct bus, but sitting down and having a smoke seemed to resolve the problem. A bus driver that didn't rapid fire an explanation at me gave me the confidence to board his bus with an understanding that if I went back to the same spot he dropped me off at, another bus would take me to the next airport when I was ready.

Finally on the way, I looked out the window at the grandeur and beauty in awe. There I was, a gal from a small town, bussing through Paris. Never in my wildest dreams did I think I would ever be there. Even with my broken heart, on my way to see the most romantic ingrained symbol of love, the Eiffel Tower, I still believed in love. So when it came into view, my hand instantly reached for the cord to get off at the next possible stop, not before getting an eyeful of the Louvre as we passed by. As the bus slowed to a stop, I heaved on my purple backpack and flew off the bus in excitement. The hours of fatigue no longer mattered, I was embracing that moment.

Regardless of the fact that I couldn't speak French, had no comprehension of the euro, and had a flight in three hours, all I cared about was the tip of the tower over the buildings in front of me. My only goal was to not get lost. So I did what I do at home, keeping track of landmarks and buildings shapes and off I went. One block up noted, two blocks over, one block around. I felt a little insecure as I lost the comfort of the bus stop but the draw to see something I dreamed of my whole adult life was moments away. As I rounded the corner, there it stood in all its glorious history. I am sure everyone has dreams of things they want to see in their lives and the possibility of that becoming real is unimaginable. So, when it does actually happen, it is very much a "holy shit" moment, as it was for me. That was a moment I would never forget, for without the painful experiences I had been through recently, I would have not been standing there. For the first time in a long time, I felt a sense of gratitude for those terrible days with Him. If this was a sample of the emotions of things to come, I was in for a journey!

After embracing all I could from the tower, I began the trek back to the bus stop. With what little time I had, I allowed myself to sit outside a small French café and experience a latte and croissant in true French style. I was getting closer to the Camino, and the excitement of the reality was coursing through me.

I arrived with success at the smaller manageable airport with minutes to spare. Quickly finding my gate, happy for my pack that didn't need to be checked, I was on board headed for Biarritz. The flight was quick and I did manage to doze a little. I was expecting the impending jetlag to arrive, so any moment I could sleep, I did. On arrival, I ventured out of the tiny airport in search of a bus that would take me to the train station.

My friend Rene, who had walked the Camino in May, assured me I would begin to see other pilgrims here, and he was right. We all looked like a bunch of misdirected turtles, ambling around with various colored shells on our backs, looking confused but attempting to give off the "I know what I am doing look." It would have taken only one of us to get on the wrong bus and we would have all followed. Fortunately, there was only one bus in the small town of Biarritz that ran through the entire route ending at the train station. For three euros, I saw my first authentic French town in its entirety.

What struck me immediately was the architecture of the buildings and bridges that have stood for centuries. Coming from my young country, Canada, I had only dreamed of seeing the history I had read so much about. The cobblestone streets narrow and winding, flower boxes on the window sills, children in traditional school uniforms, older women with bags of fresh bread with scarfs on their heads, and men milling about in cafes in fedoras. It was perfect.

Arriving at the train station, a massive stone building, I clambered off the bus with the others and filed into the station to purchase the ticket to St-Jean-Pied, the last leg of my journey to the beginning of the Camino. There was over an hour wait until the train so, unlike the others who sat in the waiting area, I went back outside to an outdoor cafe across the street. The weather was unexpectedly warm and the luxury I felt sitting there nursing my first *vino blanco* in France was delightful. I pulled out my journal and jotted down all I was feeling and observing while sipping on my two euro white wine. An older couple sitting at the next table inquired if I was on my way to the Camino.

"Yes, I am," I replied, smiling with enthusiasm.

"We too are walking this weekend, but only crossing the Pyrenees Mountains." The man shared with me. "We have done many parts of the Way, though not the whole thing at once. The Camino gets inside you like nothing you have experienced, it will drive your spirit and soul until you are home with yourself." I graciously smiled and thanked him for his words. Little did I know at the time how those words would ring true in the weeks to come.

I finished up with my wine, loaded my pack on my back and crossed the street to a small stone church. I knew there would be many churches along my path on the Camino, but I felt a strange pull to cross the threshold of that one. I am not a practicing religious person in the traditional sense, even though I was raised somewhat Catholic. I had long ago given up faith in the structure and rules of the church itself, instead, I had thoughtfully chosen, many years prior, to study and practice what I felt a true connection to, Wiccan. A belief that nature, in all its amazing growth, survival and beauty, proved to be the greatest church of all. However, at that moment I wanted to go in and say hello to God.

Maybe it was the fact the Camino is linked to a Saint, or maybe it was to ask God if he wouldn't mind keeping an eye on me and keep me safe, whatever the reason, I entered the church and felt a comfort of some sort. I took a pew and knelt on the aged leather bench whispering my private thoughts to whoever would be with me. When I stood to leave , there was an elderly priest standing quietly a foot away. As I adjusted my pack , he stepped forward and laid a hand on my shoulder. He spoke softly in French, and I had a feeling of warmth rush over me as I knew he was giving me a blessing. That spiritual moment descended on me, a light that would also be part of my Camino, the first of many.

Exiting the church, a sense of strength and spirituality engulfed my soul, not so much in a standard religious way, but a feeling of one's inner beliefs coming forward. Looking down at the small silver ring on my finger, bought shortly after the break up, the inscribed words rang clearer now, "nothing is impossible."

On the hour long train ride to St-Jean-Pied, I felt an unspoken connection to the other travelers around me, the only reason to be travelling on that train was to begin walking the Camino. Backpacks splayed on the floor, exhausted bodies dozed or stared solemnly out the window. We all had our own story for embarking on the pilgrimage. The atmosphere was quiet, apprehension of the unknown in the air. Across for me, a young couple cradled each other, sleeping peacefully, it was endearing to see such love, so full of hope.

Arriving in St-Jean-Pied, 21 hours after hugging Sarah goodbye in Vancouver, held no fanfare. Not knowing what to expect, I disembarked off the train onto a tiny platform looking at the nondescript sign acknowledging my destination. From there on I would be relying on my own power. So it began, hoisting my purple backpack on, I went in search of accommodations for the night. I could hardly absorb that I was actually in France. For all I had to endure to get there, I knew without any doubt, I would embrace all the moments in the next six weeks of my life.

Tucked into my sleeping bag in a 250 year old *albergue* (the word for pilgrim hostels), after a quick message to my children that I was safe, I felt heavy with what lay ahead of me. My thoughts, as I drifted off to sleep, were of Him. The sadness and heartbreak was still raw, but the call to the Camino was much more powerful in my heart than He would ever be again.

THE START

Morning came with the sun filtering through the small paned window at the end of the bed. I opened my eyes and focused on the view outside the frame, The Pyrenees Mountains glistened under the morning rays of sunlight. Any jet lag I may have felt evaporated at that moment for those glorious peaks beckoned me to them.

I had originally thought to spend the day in the small town before starting, but my impatience to begin was too strong. Deciding what to wear was easy, as I only had two changes of clothes to choose from.. Stuffing my pack quickly, with a double check I had everything, I descended down the tiny spiral staircase into a welcoming open room where the house hosts had laid out a traditional French breakfast. I was hungry and tucked into a freshly baked croissant, homemade jam, and cafe con leche.

"Are you traveling alone?" Felix, the house owner asked me.
"I am."
"Well then, my wife and I wish you the happiness you will find in your solitude."
"*Merci*," I replied, realizing how so very alone I was 8000 kilometers from a home I no longer had.

Felix asked to have my Camino passport so he could put my first stamp in it. The Camino passport is a document issued to each pilgrim after registering your intentions to walk through the Canadian Camino Consult. I would be collecting stamps throughout the journey as proof I had walked, and would be issued a Compostela Certificate once I arrived in Santiago.

I thanked Felix and his wife for their hospitality, strapped my belongings to my back, stepped outside the door onto the narrow cobblestoned street, and simply started walking. Being a bit of a ceremonious person, I felt the need to count my first 20 steps. Why 20? I did not know, but it gave me a menial accreditation of what I was embarking on. Those steps landed me under the small stone archway I recognized from the movie, *The Way.*, which had inspired me in the first place. Named the "St. James Gate," it was considered the official start of the Camino, and, as I was the only soul there, I took that moment to acknowledge my new beginning. At my feet embedded in the stone was a bronze shell, the sign that you were on the Camino. That symbol, along with yellow arrows, would be my directional markers for the next 800 kilometers to Santiago. The moment I stepped out from under the archway, the journey began.

The climb into the Pyrenees began far sooner than I expected. The sun was out and its warmth, which on any other day would have been welcomed, was surprisingly intense, in fact, 30 minutes in everything was surprising me.

"Shit is this uphill all the way?" I asked out loud. I processed quickly that there was a good chance my one bottle of water was not enough. I seemed to always take the bull by the horns in life, why would the Camino be any different? Periodically, I would turn around and peer down at St-Jean-Pied and see the village shrinking whilst in front of me all I saw was up.

Moments of excitement coursed through my mind for what I was actually doing. Replacing the bronze shells were the expected yellow arrows I knew to look for. They were hand-painted on random surfaces, sometimes a fence post or sometimes a rock, and would serve as my beacon to stay on the right path. The views were amazing, beyond anything I could have imagined getting to on foot. It was one thing to look at pictures on my iPad from the comfort of my couch and another to actually bear witness to real life. I climbed higher and eventually the early morning scatter of clouds above me turned into magical floating mist around me. At certain times, I could barely make out the very narrow road in front of me and suddenly a small car would appear almost knocking me off the edge.

I was surprised by my sweat and soon my t-shirt was drenched and my face dripped moisture. I also noticed, far too soon, mild complaints starting in my feet. My goal for that day was to cross the Pyrenees and reach Roncesvalles, walking a total of 24 km. I had never in my life walked that far in one go, so there was no rationalizing what the task at hand involved. I did know that once I set a goal there was no stopping me.

After eight kilometers of tedious uphill climbing, I arrived at the small checkpoint of Orisson, a stone structure that randomly appeared after a bend in the road. Its purpose was a place for pilgrims to rest, from the shock of what they had just accomplished, to eat, and, possibly, to ponder an overnight stay. It was here I found others lounging about, socializing and eating. I was hungry and very thirsty and took a seat to replenish my confused body. Having started around 8 am, it had taken me two and a half hours to walk eight kilometers. While munching down on a ham baguette, I calculated I was covering just under four kilometers an hour, in my mind, that wasn't too bad. Then it hit me, I would be doing the same thing for the next four or five hours! I let that thought slip away quickly as it scared the shit out of me. Suddenly, without any training, it seemed ridiculous to be doing what I was doing. But my attitude towards life had always been to just do it, no matter how har. It was an inner strength I always had.

Back on my feet, I started the next 16 km. There would be nothing but a mountain between where I stood and Roncesvalles, so the continual uphill commitment was ahead of me. My legs began to adapt to swinging one in front of the other, like a pendulum, without much thought. I no longer looked behind me but strained to focus on what was to come—ah, just like life should be.

The small road turned into a nasty, gravel trail that challenged both my balance and fatigued legs. The incline remained ever so steady without any relief. I found myself leaning forward over my hips so I could dig my feet in more and compensate for my, seemingly, heavier pack. The trail finally gave way to more forgiving footing, which led into an alpine meadow but despair etched its way into me for I looked ahead and all I saw was up, a whole lot more of up. There was no way to gauge how far I had yet to go. I had not seen another soul since Orisson and my emotional state was being challenged.

"What the fuck am I doing here? How much farther up can I possibly go?" I called out to nothing. I just bowed my head and continued on, becoming oblivious to thoughts and surroundings, there was nothing else I could do.

I am a pretty tough chick and can handle more things than most people, but, honestly, the unknown was kicking my ass. Tears sprung to my eyes as I thought of Him. It was His fault I was broken emotionally, sore in every part of my body on a mountain somewhere between France and Spain, doing something so far removed from my normal life. I was not ready to feel any gratitude yet for being in that magnificent place. Then, as if it knew I needed its presence, I glanced to my left and, standing on a knoll quietly, making eye contact with me, stood a horse silhouetted against the bright blue sky. Being a passionate, professional horsewoman for more years then I could remember, the presence of that creature gave me the nudge of hope I so desperately needed.

Around a bend in the trail, I came upon an area that overlooked the expanse of an amazing view where several other people sat on random logs all looking as lost as me in their own thoughts. Nobody spoke. I unloaded my purple monster pack to the ground and slipped out of my boots and socks to let my feet breath and feel the sun. I could then relish for a moment in the beauty of the heights I had climbed. "Being in the mountains" took on new meaning for me. There was definite energy amongst all of us at the rest stop, and I had a feeling of accomplishment.

In my hours of research of the Camino, I read that several people had died up in the Pyrenees Mountains and I had seen a few markers along the way that confirmed this. I sure as shit was not going to be one of them, I laughed inside. Clambering back into my gear I carried on, bent at the waist, as I continued the uphill climb on the very narrow rocky path, it just wouldn't relent. My right ankle, reconstructed in a surgery 20 years ago after an accident, reared its ugly head. My left ankle did all it could to compensate for the pain by carrying as much of me as it could. I felt anger swirling in me, how stupid could I have been to hop a plane across the world and walk up a big fucking mountain. As in life, sometimes I didn't always think things through.

I eventually reached the height of 1800 meters and finally, far below, caught a glimpse of my destination, Roncesvalles, looking much like an oasis in the desert, was in sight, the end was near.

I had been walking for seven hours and I felt renewed as I began the descent but negotiating the trail was not easy on legs that felt like noodles. I had thought going down would bring relief but I was so wrong. Along with my sore back and angry right ankle, my knees decided to join the party. The change in angles was not serving my body any relief. Keeping my focus on the minuscule view of the church steeple of Roncesvalles, I carried on. An hour and a half later, the earth finally gave way to the first level ground since I walked out of the *albergue* in St-Jean-Pied.

I walked along an enchanting forest path leading into town and my spirit started to lift. The yellow arrows guided the way and I received the warmest welcome of four horses walking towards me, loose with their own path to follow. Entering through the backside of the town I passed an ancient monastery built of large stone blocks with moss draping along its sides. Many would opt to stay there for the night in the large dark cold room that could house upwards of fifty beds, but for me, after what I had done I wanted the comfort of my own room. Being alone with myself was what I needed.

Learning to communicate in traditional Spanish became the next task. I was thankful for my interest in the language, acquired from several trips to Mexico. I succeeded in securing a key and a glass of *vino blanco* and, before I went upstairs, I sat outside in the warmth of the setting sun and reflected on the day's walk. At some point, I had crossed over from France into Spain, apparently that crossing was a cattle guard I could not even remember doing some six kilometers prior. I had climbed into the clouds, seen views of beauty, pushed my body to do the unthinkable, and challenged my spirit beyond measure. And it was only day one…

I gratefully sign on to the *albergue* WiFi, sending a blanket message to my kids that I was alive and well. I also sent a personal message to Rene, my 1-800 "how to do the Camino" friend simply stating, "Hey, so you told me the walk through the Pyrenees would be like walking up one of our local ski hills in B.C., I'm calling bullshit!" attaching an emoji of laughing with tears. As luck would have it, he responded immediately, "If I told you how difficult it was, you may never have gone." I smiled at the simple explanation and knew he had done me a huge favor.

Having drained the last of my wine, I felt the weight of my body, exhaustion was coming through in waves, I needed sleep. Climbing the stairs to my room put the final drain on my 5'6," 120-pound frame. I hardly noticed the surroundings of my refuge for the night as I stripped off my backpack and boots and climbed into the bed, filthy stinky clothes and all. There was nothing left but to drift off to sleep. 790 kilometers to go.

REALITY HITS

Searing pain bolted down my right leg the next morning, waking me from the nightmare I was in with such force I sat up too fast and got dizzy. While I rubbed my leg in an attempt to soothe the pain, the coughing began. Deep within my chest, it bulldozed its way through me. Confused, between the pain in my leg and ankle and the convulsive hacking, I forgot where I was. Initially hell came to mind. The realization hit me hard and fast as I attempted to get out of bed to do something to get away from the pain and only succeeded in falling down on my knees on the hard floor.

"This can't be happening to me, not now, not here!" I cried out to the empty room. For years in my career, I had continually pushed my "bad" ankle training horses professionally, but the pain that day was beyond anything I'd ever experienced. I used the bed and a chair and pulled myself upright only to discover in horror I could not put any weight on my right foot. With one leg and the aid of the wall, I hopped towards the tiny bathroom. Grabbing a hold of the sink, I looked into the mirror and what reflected was a dust-caked face, swollen red eyes, and exhaustion. Then the coughing began again, I hung onto that sink for dear life as my body heaved through the need to use every muscle to expel, all on one leg. I was a pathetic, fucking mess.

Once the coughing subsided, I eyed the tiny shower and knew I had to get in it, clothes and all. I tentatively put weight on the ball of my foot, which allowed me to hobble into the cubical. Over the years I knew if I got the joint moving again I would be okay, it was all I could do to get my body to get its shit together. Under the spray of the lukewarm water, I commanded my mind to overcome the pain and for the sake of sanity to stop the horrible cold that was invading. Great, so this was day two. I stumbled around my room and used every upright object for support as I got dressed, pinned wet clothes to my pack, hoisted it on my back and began the long journey downstairs.

My ankle was three times the size it usually was and the joint was protesting against the barrier of swelling. There was no freedom of movement in my ankle as I gazed down a flight of stairs on the ball of my foot. The only way to get down was on my butt or go down backwards, I chose the latter.

"Can I help you?" offered a man as he passed me.

"Thanks, I'm good," I replied in my independent, crusty voice. Limping into the communal breakfast room, all eyes were on me and the pity I witnessed embarrassed me. It appeared no one else was suffering after day one, I was pitiful. I sat down to a Camino breakfast of porridge and fruit and was quickly questioned by the others at the large table.

"Do you need pain pills?"
"Would you like some cold medication?"
"Foot massage?"
The concern I felt from these strangers was endearing and brought a solitary tear to my eye.

"I appreciate you all, but I will be fine, the tough person that I am. This ankle has been my enemy for many years. In some way, we are old friends who don't always get along." The table erupted in smiles and laughter, which was all I really needed then.

After breakfast everyone stood to start their day but I waited behind for a few minutes to carefully stand with no eyes watching. History dictated that when my ankle was inflamed, the first few steps after sitting again were not a comfortable process. Cautiously I stood up, limped over to my backpack, and headed outside into the shining sun and refreshing crispness of the morning air. I plunked down on the steps and pulled out my pack of smokes and enjoyed every draw from that first cigarette. It was my last pack and I planned to quit when it was finished, but, right then, it was the comfort I needed. Not for a moment had I contemplated not walking, regardless of the pain and the existing cold.

Across the street from where I sat was the municipal road sign I had seen in the movie, *The Way*. It read, Santiago 790 km. I limped over to stand beneath it feeling a sense of awe that there I was and it was all I needed to turn towards the path and follow the arrows. A few pilgrims passed me by at great speed as I stumbled along, wishing me "Buen Camino," translated as Good Camino, a phrase I would hear and share many times over the journey. As alone as I felt, it was a comfort to hear their words.

The path was flatter and kinder compared to the previous day, and the more I walked the more my ankle loosened up, though still uncomfortable it was becoming bearable. I was grateful for the bottle of Advil tucked into my waist pack, it was certainly helping enough that I was able to focus on the beautiful forest I was walking through. I wondered how many had walked the same path. Here I was a gal from Canada, part of that history.

After the passing of the morning rush of pilgrims, I was alone again and it felt daunting yet peaceful. That day the walk was very different from the Pyrenees, with gentle slopes and quaint, little villages to wander through. The locals in each would acknowledge me with a nod or a smile, which made me feel like an honored pilgrim. The weather was lovely, apparently unusual for Spain, with warmth from the sun and brilliant blue skies. Butterflies flew around me, which made me think of my mom who had passed away two years earlier, it was a gift of comfort. I thought about my capacity to love and care for those who had crossed my path throughout my life, though it had not always worked out for me. There were times when my world was less than honest and the pretense from others would rule my heart, only because I would adapt to them and situations as an ingrained pleaser, at my own emotional expense. Often I thought I should be more blunt, more in command of my own voice, but that just hadn't been my style. Out there, alone on the Camino, simply walking, I knew I had to change the way I appreciated myself. Geez, "love myself," how many self-help books had I read that in?

Around the 18 kilometer mark (when did that happen?), I came out of a dense forest to a road where a small caravan was parked and offering refreshments. What luck! I took the break with a yummy cafe con leche, some biscuits, and a smoke. (I love when things happen out of nowhere.) A cardboard box beside the caravan was filled with a rainbow of bras and underwear with a sign reading, "Donate either item and you will find love on the Camino." I had no desire to donate, but instead found a quiet spot under a tree to enjoy my treats. Once I was somewhat rested, I stood up but my damn ankle had locked up again with a distinct new pain in my right hip socket. "Christ on a cupcake! Really!...Really!" An inconvenience, but onward I went back into the forest.

The last three kilometers of the day would bring me to Zubiri, marking the total day's walk at 21 kilometers. I had no *albergues* booked ahead, so each day was an adventure of finding one. The sun was hot by then and sweat had pooled at my lower back and in my boots. It was a long steady descent, which invited my neglected knees to feel uncomfortable again, along with the ankle and hip pain and still the stupid cough—all I could do was laugh at myself. I arrived in the small town of Zuburi with a focus to find a place to call home for the night. The first one I came across looked good enough for me, so I entered the four story upright stone structure. Of course, I then had to climb three flights of stairs to get to the reception area. I arrived at the end of the mountain of stairs, and found a small desk against the wall, behind it sat a tiny, older man who looked part of the structure, like he had always been there. He greeted me with a warm smile and a wave of his hand to come closer.

"*Hola peregrine, vein a descansar un rato?*" Loosely translated, "Hello pilgrim, come rest awhile." That lovely soul handed me a rather large, antique key for about $7 Canadian and pointed to the floor above us. I climbed the additional set of stairs and found a small room consisting of two bunk beds that I would share with three other people. They were all home and, upon entering, I immediately recognized the two young lovers from the train, who reached out to me with an introduction, Olga and Jaro. The third roommate, a mature man with a Belgium accent, introduced himself as Jos. The bunk available was above him as he pointed out to me and I gladly heaved my pack on it. A quick shower in the floor-shared bathroom washed away the caked on snot and dirt from both me and my clothes. With two Advils, a cotton sundress on, backpack safely put away, and clothes hanging from my bunk to dry, I headed out to find supper.

Not far from the *albergue,* I found an outdoor cafe with other pilgrims and settled at a table for one. I ordered a glass of *vino blanco* and a fish plate that was offered on the specials board.

"Hi there, are you Canadian?" a woman asked from the table beside me.
"Yes I am, and I am assuming you too?" I replied smiling.
"Yup, I am from Toronto." She reached out her hand, "Dunja."
"Great to meet another Canadian, Dunja. I am Tess."

This was the first real conversation for me since I'd left Vancouver. Dunja explained she too was traveling alone and how she had been ill with pneumonia and had to make a deal with her husband to be able to still go on this trip after her recovery.

"I have my pack transferred from town to town by a service," (WHAT! You can do this?) she explained, "and *albergues* are not an option as they are too damp and cold, so I must stay in proper hotels."

It was easy to understand that her husband loved her dearly and wanted her health to be a priority, especially as both of them were in the medical field, he a doctor and she a nurse. Dunja continued to share that when she finished the Camino she would fly to Paris to meet with her husband, where they would celebrate their anniversary. Dunja and I became instant friends, sharing stories of Canada and our Caminos. I enjoyed hearing about the glorious marriage she and husband shared after 30 years and realized privately that He could never give me that.

Then Eva came along, a robust Dutch woman who pulled up a chair and joined the two of us. I commented on her Dutch accent right away, explaining I was of Dutch descent, which engaged us in a brief exchange in our native tongue. Dunja laughed out loud exclaiming she was feeling left out. Having been on my own for the last three days, I enjoyed the company of these two independent women. Eva shared with us her remarkable story:

"I walked out the door of my home in Holland two months ago, leaving a very supportive husband in charge of our two children. I explained to him I needed to go do this and he pretty much shoved me out the door. I have been walking since, it has been such a journey." They too were planning to meet, but different from Dunja's plan, Eva's husband was going to join her on the last leg of the Camino into Santiago in five weeks. I was in awe sitting with these two amazing women with their heartfelt stories. My thoughts wandered to Him, wishing He'd had the capacity to love me the same way. They both asked me my story and I mumbled something about a long time bucket list, which was not true, but I didn't feel the need to share my truth amongst all the good love vibes going around. We sat together laughing and chatting until the air chilled. My shoulders started to ache and fatigue engulfed me, so I bid the ladies good night, with hopes we would meet again, and headed to my bed.

Olga and Jaro were sound asleep when I got back to the room, and Jos had yet to return. I crawled up into my bunk with my iPhone, connected to the WiFi, and reached out to Rene.

"Hey, I am hurting physically beyond anything I could have imagined," I sent him. Rene was online and promptly responding with, " This my friend is your new normal." Rolling my eyes in the dark I responded with simply, "Great…"

I also touched base with a friend I had recently met before leaving who had the kindest of hearts, James. He had been a great ear to listen to me talk about my upcoming adventure before leaving and always appeared genuinely interested. James was sad to see me leave, the time we had known each other was short, but a connection had been made and I cared enough to send a note that I was conquering the kilometers.

Then I sent a message to Him. Reasons I could not possibly understand other than to band-aid the pain that remained in my heart. He did not reply.

Tomorrow I would walk to Pamplona, famous for the running of the bulls. I was looking forward to seeing the first city on my walk and was contemplating a rest day there. Curling up in my sleeping bag, I drifted off into a night of restless sleep.

THE UGLY WAY

Waking up in the middle of the night had not been the plan, but pain did that. My shoulders and legs were cramping as I lay trapped in my top bunk in bewilderment. The three other people in the room snored peacefully in their beds. Quietly as I could, I reached into my pack for muscle cream and melatonin to try and knock myself out again. I succeeded in finding both in the dark of the room laid back down in hope sleep would return.

Dozing off and on, I never really found rest again, so I gave up and, in silence, grabbed my things and slithered off the top bunk. Out in the unlit hall, I stuffed my pack, brushed my teeth and headed down the stairs backwards again. In the kitchen below, the 20-foot harvest table was laid with breakfast consisting of a large baguettes, jam, and fruit. It was the nourishment I would rely upon till lunch. I ate alone in the stillness of the early morning start, it was a solemn way to start the day.

Once outside in the darkness, strapping my purple pack to my body, I began a feeble hobble when an Australian couple approached me.

"Are you okay?" the woman asked.
"I will be once I get going. I have an old injury that is voicing its opinion about this Camino thing," I responded, smiling.
"Right then, hang on a minute," the woman dropped her pack and began zipping and unzipping the many compartments. "I have these super effective anti-inflammatories from back home." Her husband confirmed they would do the trick. "Here you go, take the whole strip," the kind woman offered.
"Thank you, thank you, thank you," clasping my hands together, "you are very kind." Never anywhere would I have accepted random pills from a

20

stranger, but, out there, it was a gift given in trust. I watched them disappear into the dark as I popped two of the little gems with a slug of water and started the 25 kilometer journey to Pamplona.

The night sky was slowly giving way to the morning light as I picked my way through the forest path. Coming out of the woods, my eyes were assaulted with what I could only describe as a wasteland. For all the beauty I had encountered so far, the view was now violated. It was barren, void of trees, with garbage strewn about and abandoned factories in various stages of collapse. My memory triggered reading about this place in my research, and it was rightly dubbed, "The ugliest place on the Camino." The saving grace, as there is always some beauty in everything, was within the enclosed barbwire fence I had been walking along was a small, white horse. It promptly trotted up to the fence line to say hello. I stopped to greet the beautiful creature, breathing in the horse smell I knew so well as he nuzzled my fingers for comfort in that strange, foreboding place. The beauty of that moment was more memorable than the ugliness that surrounded us. Saying goodbye to the angelic creature, the path led back into the dark forest but not before clambering over blocks of broken concrete. Regardless of the sun trying to shine, within the grove of trees, there was still a chill and dampness in the air. I had anticipated the warmth of the previous day far too early and I was cold. My arms felt like icicles while the sweat from the pack on my lower back turned cold quickly.

A few other pilgrims in small groups passed me and we would all acknowledge each other with a *"Buen Camino."* Other than those brief encounters, I walked in solitary for several hours. Being alone with your thoughts, with nothing to distract you, takes you down a realm of some deep thinking. I acknowledged that though walking alone is harder it also makes you stronger. Someday, that would make sense to me but for survival on that day, I laughed through my tears and reminded myself that everything happens for a reason. Then from behind, I heard my name being called out.

"Tess! My fellow Dutch pilgrim, how are you?" Eva called out, my new friend from the previous night.
"Hey, Hi Eva. I am doing pretty good taking one step at a time," I replied, as she fell into stride with me.

Eva suggested I try her walking sticks for a while, with explicit instructions on how effective they would be when used properly. I had never been the most graceful of movers and found them to be more awkward than useful after a kilometer and offered them back to her.

"Eva, I feel like a crab trying to walk in a mudslide!"

"Ah that's funny Tess, true it is not for everyone. You are a strong woman and you will find a way that works best for you."

I assured her I would as we bid goodbye and off she marched at her robust fast pace, walking sticks-a-flying. I was amazed by the connection that ignited amongst strangers on the Camino as we all traveled with the same goal.

The hours passed by putting one foot in front of the other while listening to the music I had thoughtfully loaded on my iPhone. Hunger was gnawing at my stomach, reminding me that my body needed fuel. I had never been a big eater, nor gave conscious thought to food but never before had I been somewhere where I could not get some, until then. The fact that I had not thought to buy snacks for the road was quickly becoming a lesson I needed to learn. As I rounded a bend in the woods, I came upon a man with a blanket laid out with fruit, nuts, and water on it, a gift from the Camino. He offered me a rock to sit on and encouraged me to take whatever I needed from his offerings. Relief came over me in waves at how that gift appeared when I needed it most. I delighted in the small feast of an apple, a handful of nuts, and water, thanking the man with a small donation.

The closer I got to Pamplona the more often I would see hamlets popping up out of the countryside. How I loved to look at the architecture of days gone by and appreciate its history. The smiling faces from the residences offered me a sense of pride as a Camino pilgrim. I stopped in one hamlet, Viscarret, for a coffee and a chance to rejuvenate. I sat on the grass with the cafe con leche, removing my boots and socks to feel the freedom my feet needed. My ankle was doing better, thanks to the anti-inflammatories given to me that morning, and I had begun to understand and accept the complaints my body was telling me. I knew I would start to pay better attention. The sun was warm and I appreciated the time of rest with my coffee. I pulled out my journal and jotted down thoughts, which evoked a couple of people to ask me if I was writing a book. I smiled and explained I was just journaling my experiences so I would never forget.

Eight and a half hours after leaving Zubiri, I arrived at the gates of old Pamplona. Having spent the last few days in quiet villages, I entered into a city surrounded by people, bikes, cars, and tons of activity. Famous for the running of the bulls, I was walking down the same cobblestone street where the yearly event occurred. Similar to the excitement of seeing the Eiffel Tower, I was once again amazed to be in a place I had only read about. I found myself in front of a large cafe with outdoor tables so I shrugged off my pack and absorbed my surroundings. I ordered a cold beer and symbolically pulled out a well-deserved cigarette from my newly purchased Marlboro pack (I wasn't ready to quit yet). As I reveled in the surroundings, two young Irish girls at the next table asked me where I was from.

"Canada," I replied.
"Really? You look so very European," one girl responded.
"I do? I have never really thought of it before. I am from Dutch heritage so maybe that is why," I laughed through my response.
"That would explain your look then. You certainly look like you belong here."
"What a lovely thing to say, thank you," I said.

That exchange got me thinking about how I never really fit in at home with my looks. We as women cannot help but compare ourselves, not only with others, but also with how society dictates we should look. I had never really thought of myself as an attractive person but there, in Spain, I made the connection. The high cheekbones, deep blue eyes, and strawberry blonde hair made up the appearance of who I was, and who I was was European. The short conversation with those girls helped me to see the beauty in myself that I had not seen before, a gift from the Camino.

The sun had begun to set and I thought I better get a room for the night soon. Downloaded on my phone, thanks to Sarah, I had an app that would show me the *albergues* in the area. I found one located not far from where I was and set out, following the blue dot on my screen. I still managed to miss a turn, which resulted in me standing in front of a shoe store with a Camino shell in the window. My feet found the way over the threshold in search of what? I wasn't exactly sure. A small character of a man approached, smiling, and welcomed me with a warm hug, I was bemused. He looked down at my feet and declared in broken English, "We can do better than that." There is always a reason you miss a turn in life and it usually leads you to a better place. For a mere 50 euros ($74 Canadian), I walked out of the store with insoles that felt like clouds, cushioned socks, a new waist pack (the thrift store find from home was already falling apart), and a walking cane that I decided to call Harold. The proprietor wished me *Buen Camino* and sent me in the right direction to get to the *albergue*.

I knocked on a small, nondescript wood door and was let into an old building and showed to my room. I promptly dropped my pack on the narrow bed and set out to explore the amazing city I was in. It struck me funny that I could walk all day over many kilometers and still give in to the curiosity of discovering a new place rather than rest. Nothing could have stopped me from venturing out, especially without the purple monster of a pack.

Teenagers talked amongst each other, arms gesturing every thought, women biked past in pencil skirts and high heels, men walked six dogs on one leash, families of many generations sat together on benches, and lovers kissed passionately on balconies—I embraced it all. Not one person had their face glued to an electronic device like I was accustomed to back home.

I walked the narrow streets, one after another, and came across the bronze statue dedicated to the running of the bulls. It was almost half a block in length and depicted the anger and frustration of the animals and the adrenaline and fear of the runners. There was so much power in that stationary statue that evoked an emotion I was sure the artist had hoped to convey. At that moment, the Camino became so much more than a walk across Spain, it became a chance to learn and feel the history and experience connections with like-minded people. It was then I went to purchase the Camino shell I would attach to my pack, the honored symbol carried by a pilgrim.

Shortly after I ran into Olga and Jaro, the young couple I seemed to keep finding. We embraced like old friends and marveled at finding each other again. We shared our experiences of the day with all the laughter that came with it. I learned from them that they only had a week to walk, for they would be moving to England from Russia to new jobs. They were taking the shorter version of the Camino as a holiday before their upcoming hectic schedules began. I enjoyed hearing the different reasons why people walked the Camino, it varied from health to healing. For me, I was no longer sure because my thoughts had drifted since I decided to do this journey months before. When the adventure began, a thought had been whispered into my broken heart: let the journey unfold, let it be magical, let it be mine, the hard work is behind me and I am going to reach my reward.

Later that night in my room, after washing my clothes, showering, popping a few blisters and massaging my legs and ankle, I lay my head on the pillow. I was out in a moment to relive in my dreams the sadness that brought me to the Camino.

THE TEARS

The following morning I found the only open cafe nearby. It was a dreary wet morning and the street lights were shedding the last of their nightly glow. I sat at one of only two tables outside and watched the other pilgrims start their day draped in huge ponchos over their packs. With all the colors of the rainbow filing down the street, they looked like a flock of exotic birds and it made me giggle. I was humbled though because I had not yet purchased the recommended poncho, instead, I enjoyed the warmth from my coffee and Marlboro and waited out the rain, the Tess way.

I reflected on the glorious evening the night before when I explored the city for hours. I loved the way people sat on curbs drinking and socializing wherever suited. Dogs, of every size and breed, seemed to be owned by everyone. The streets were so narrow I could not imagine a car driving on, yet on occasion, I had to plaster myself against a wall to not lose any body parts. In three short days, I was already embracing so much and was keenly looking forward to all the experiences and adventures that were still sure to arise.

The rain let up after I finished my coffee, which signaled the start of the 25 kilometer walk to Puenta la Reina. I was using the Camino app on my phone to help me determine how far I would walk and where I would try and be at the end of the day. As much as I relished in the adventure of winging it, it seemed like a good idea to have a little guidance. The biggest concern I had then was not being able to find a bed at the end of the day. I had heard stories of pilgrims arriving late in some of the smaller places with there not being any beds available and having to continue walking. I could not imagine clocking in so many kilometers and then not being able to stop when the body was done.

When I was outside the city limits of Pamplona, Dunja, my other friend from Zubiri fell into stride with me. Having remembered our enjoyable meeting, we comfortably walked a short time together, not saying much. It was kind of funny looking at Dunja in her pretty matching lime green attire—how fresh and clean looked— and then there was me, the poster girl for a thrift store. I remembered her Camino was different from mine and refrained from judgment because it would have come from jealousy. We were all engaged in the journey, just like life, for all our own reasons. That was a lesson I would not forget, a lesson to refrain from judging others, as we are all just doing the best we can with our lives, whatever the unseen story is. When Dunja and I parted ways, I hugged her, thanking her for the lesson quietly with a hug.

I followed the faithful arrows that appeared once again where the sidewalk ended. They directed me away from civilization and on to a path that took me up a small mountain. My ankle was being cooperative, considering the climb consisted of large rocks on top of washed-out footing. I had to carefully place my footfalls and was grateful for the little old man who insisted that I purchase Harold, my walking cane, as he proved his worth by keeping me balanced. The heat from the sun was already intense that morning and the familiar dampness under my pack returned. Spain had experienced a far warmer autumn than expected and, with the rain, the humidity was high. Being a lover of the sun my entire life, I felt gratitude in its comfort. After a couple hours, I reached a hillside that had been recently harvested. The field had been sheared to its roots, which allowed me to distinguish the path meandering its way to the mountain crest. I was amazed at being able to see the simplicity of the Way ahead, it was a personal invitation. As I followed the path, a young girl appeared out of nowhere and stopped me with her radiant smile.

"Hello, how is your Camino today?" she asked in clear English.
"It is a good day, thank you," I replied.
"Do you have a destination for the night?"
"I do, I am planning to get to Punta la Reina by nightfall," I replied.
"Well, at least that is my plan."

She smiled, "My family has a small *albergue* just before that village that we have only just opened up. Would you like to stay there?" she asked." It is very inexpensive. We have cold wine, good food, and a very nice garden."
"Gawd yes, that sounds perfect, especially the wine." We both laughed. She handed me a small hand-drawn map and assured me that I would be welcomed on my arrival. I thanked her, feeling the relief that I had my

lodgings sorted out for that night, and out there on a mountain path with no electronic conveniences.

When I arrived at the top of the climb, there stood a lone, beautiful tree draped in fall colors with an inviting bench underneath. A beautiful woman sat on the bench and smiled at me as I came closer, I took a seat alongside her.

"Amazing view from here, so worth the climb," I said.
"That it is," she replied. "Where have you traveled from?"
"Home is Canada, what about you," I asked the friendly pilgrim. " I am from Brazil," she said, reaching out her hand to me. "My name is Chris."
"Nice to meet you, Chris, I am Tess."

We sat in silence for a while taking in the clear air and beautiful view, secretly resting our sore feet. Chris broke the silence happily sharing her story with me. I appreciated the openness of people on the Camino, or, rather, their need to talk to someone on the journey. Chris told me her husband noticed she was getting stagnant in her work as an artist and encouraged her to walk the Camino to reboot her passion. He helped her plan her journey, assured their two young children she would be fine, and sent her on her way. Chris spoke with such love and appreciation for her husband that I could not help but be a bit envious of the love the two shared.

It made me think of the relationship I had had with Him, so wishing he could have been that someone. Why did the demons of his past shroud His heart in a way that he could not see how much I would have continued to care about Him?

"Tess, why do you look so sad?" Chris asked, interrupting my thoughts. "I see that amazing turquoise dragonfly tattoo on your arm, the same color as your eyes that need to shine to show the world it is a good place."

I was moved by her simple words, even though she did not know my story. It helped me step further inside myself and see more of who I could be. I needed to walk again. As I bid a warm goodbye to the beautiful soul on the bench, I walked away with an appreciation of those that came into our lives for a reason.

Why was I walking the Camino? I remembered reading about how there was a time when the Camino was used as a pilgrimage for criminals to make penance for crimes they had committed. My own crime was becoming clear: I had spent most of my adult life not believing in myself enough to see my own worth. Regardless of the many great things I had accomplished, all I could see was the failure in the things I could not achieve. As I was becoming more aware of it, all I wanted was to undo the wrong I had done to myself. The Camino, my Camino, was going to be my opportunity for penance, to find a way out of always feeling less than good enough. I made a commitment that day to be kinder to myself and shed negative thoughts that did not serve me.

Atop the next peak was a monument depicting pilgrims overlooking the valley and mountain I had just walked up. I had looked forward to seeing it, as it was one of the landmarks from the movie, *The Way*, and ,as I approached, my steps quickened in anticipation. But what awaited me, shocked me. I had expected a peaceful, almost sanctuary like, atmosphere.What I found was a tourist destination. Yes, there were pilgrims, but there was also many people arriving on buses! I was thoroughly disappointed and had an overwhelming urge to leave as soon as I arrived. The spiritual place in my mind did not find the Zen I had visualized. I snapped a quick photo of the monument and hurried through the crowds to the path that would take me off the hill. The terrain going down was made up of small, round rocks that made any quick movement impossible. I worked on letting go of the disappointment, instead, focusing on placing my feet so not to fall.

I managed to arrive safely at the bottom and there in front of me was a large pile of rocks shaped like a miniature mountain. I had read to bring special rocks from your homeland with you ("you will know what to do with them"), so I had secured on my backpack a small leather pouch with a collection of carefully chosen rocks, with a few from close friends. Until then, I did not know the purpose, but standing there in front of the handmade monument I knew. An older man was standing beside me and I asked him if he wouldn't mind untying the rock satchel from my pack.

"Yes, of course," he replied in a long-drawn breath. He handed me the satchel and I reached in and withdrew a red rock that I had found in Sedona, Arizona. I placed it on the pile with the others and felt the symbolism behind the ritual as my first gift back to the Camino.

A few kilometers after leaving the rocks I realized I had lost my beloved purple scarf, my favorite purple scarf I had for years. I promptly turned around to start retracing my steps to find it, only to stop myself, remembering something Dunja had told me the night we met in Zubiri. "Never retrieve something you have lost, for it is gone with reason." I was not sure why that phrase stuck with me, but then, it made me realize how I could love something material to the point of going back to a place of such disappointment. I didn't want that for myself anymore. I turned back around towards the Camino, and went forward.

Nearing the day's end, I was ambling across a large field void of any new growth. It was there I was suddenly overwhelmed and abruptly dropped to my knees submerged with grief. I cried the tears I had not yet. I cried for the loss of the life I would never have with Him and cried for the empty ache in my heart. The emotions of the lessons I learned throughout the day engulfed me in one fell swoop. I was being forced to deal with the thoughts that flooded my mind, and in doing so, began the process of true grieving, which would allow the healing. That release gave me exactly what I needed. Exhausted, I staggered to my feet and walked the remaining three kilometers accepting I just had the breakthrough I needed to begin recovering in order to forgive myself.

JUDGEMENT

Finally coming into the yard of the *albergue* was an emotional and physical relief after the eight hours of not only walking but also intense inner journeying. I was greeted, as promised by the young girl, by a friendly host standing on the porch.

"Welcome, *Peregrino,* welcome." I climbed the steps and he reached out immediately for my purple monster knowing my back was sore at that point.
"Thank you, it has been a long day." I acknowledged his kindness. Arnie, as he introduced himself, assured me this section of the Camino usually was for all.
"Please sit here on the porch, I will bring you a *vino, blanco* or *tinto?*"
"*Blanco, mucho gracias, Arnie.*" I smiled in gratitude.

Turning to the inviting porch, I saw a big table set with other pilgrims gathered, drinking wine in reds and whites. I passed by the group smiling and scoped out a quiet spot at the far end of the porch. I took my place on the wide cement railing with my back against the sun-warmed post and stretched my legs happily in front of me. Slowly sliding my boots and socks off, I felt the instant relief of the air touching my bare feet, I had never appreciated my feet more. The inspection of my feet revealed an array of blisters cresting each and every toe, it was a hell of a display. I laughed and pondered how I had not felt them. My painful ankle, however, did not offer me any humor. It was grossly swollen and bursting against the Elastoplast wrap I had put on that morning for support. I removed the wrap and elevated it, it was all I could do. My *vino blanco* arrived and I put my ankle thoughts aside as I sipped on the medicinally refreshing nectar. The cold wine washed passed my parched mouth, down my dry throat and nursed my soul.

Sitting in my solitude, not too far from the table of pilgrims, I could not help overhearing the conversation. Everyone was speaking English, though each person spoke with an accent for their own respective country. Jos, the bunkmate from Zubiri, held court within the group. I listened and learned he was a psychiatrist from Germany and was questioning a middle-aged woman from Canada about her reasons for walking the Camino. It was obvious she was not comfortable with the free analyzing and was desperately defending herself. Jos was being very judgmental, which I thought was strangely opposite to what I would expect from someone in his profession. He was adamant with his opinion on her thoughts and her reaction to insecurity and fear was obvious. How words can cause such pain. It was a good lesson for me to be more aware of my own words, at the expense of the poor Canadian. Chris, the Brazilian beauty from the bench earlier that day, chirped in to challenge Jos on his scrutiny. A loving looking couple also at the table listened intently. The look on the woman's face was full of empathy and kindness as she reached over to lay a hand on the victim's shoulder for comfort. It is what I would have done and I instantly liked her.

Arnie came out on the porch then and announced dinner was ready. The Canadian woman was the first to jump up and we all followed with relief, the spell of Jos had been broken. Everyone ahead chose their seats at the large table and I sat at a single chair next to the couple I had just admired. The three of us were quick to engage in a conversation full of good energy and a connection was felt immediately. Carlota and Jorge, I quickly learned, lived in Spain. They had become a couple recently and Carlota, who had walked the Camino once before, had suggested to Jorge they walk it together. I was intrigued by this as I expressed what an excellent way to begin a relationship. They both giggled and agreed on the very awareness of the growth and challenges they would face.

"Well, I would think if you survive this, the two of you together will be good to go." I muffled through fits of laughter. Jorge joined in my amusement, claiming he was a strong-willed Leo that was learning to take a bit of a backseat to Carlota to "lead the way." I smiled in response, being a Leo myself, I could appreciate how that can be difficult with our type-A personalities. We spent the rest of dinner sharing jokes about strong women, sore feet, and poor Leo men. I was so happy to have met those two, and the bond we all felt was fantastic.

With hugs and cheek kisses, good nights and assumed goodbyes, we ended the wonderful evening. Crawling into my bunk, I felt the warmth of happiness from having met the two kindred souls, and wondered if our paths would cross again. That night I slept peacefully.

WASN'T THAT A PARTY

The Way of St. James, The Camino, is regarded as being separated into three stages that inevitably affect every person who chooses to walk it. The first ten days are considered a physical awareness, to push your body to levels of discomfort and unique pain not thought possible. The next ten days bring you to a level of acute awareness of your thoughts, which could begin to flow from a beginning to an end. The last ten days is the spiritual connection of all you may have learned and what you have overcome. It is in this stage you think you could never leave that place and return back to the normal world.

At the start of day number five, I enjoyed the warmth from the sun cascading over my shoulders, it was comforting. I allowed myself for that moment to forget about the blisters, the sore ankle, and Him, and instead plugged into my music and walked to the rhythm of each song. I passed through tiny villages, admiring the stone dwellings, smiled at the livestock grazing, and enjoyed the changing of the leaves, it all held magic for me. A serenity filled my senses that could only be found when traveling by foot, for it allowed you to see things easily missed in our hurried lives.

Leaving the comfort of civilization, I was again walking the Camino on uneven ground and faced with a hill so steep, I had to use my hands on the ground to pull myself up. The purple monster felt heavier that day. I resented the weight of my life inside it for it reminded me of the weight we pack around in our minds and how it affects us. Knowing I would eventually remove my pack at the next rest stop for relief, I was thinking of how I might do that with thoughts that no longer served me. We all have the burden of despondency roaming around in our heads, problems we cannot solve, pain others have caused us, and our disappointment in ourselves. Like the purple monster, we need to take a rest from the weight every once and a while in order to be stronger to carry on.

I arrived at the top of the excruciating hill, soaked in sweat and breathless, pleased to find another pilgrim shrine. I reached into my pouch and retrieved a rock from Zion Canyon, Utah, and placed it amongst the others, acknowledging yet another mountain accomplished both physically and mentally. The path then began a steep descent into a small village that welcomed me with a place to rest and enjoy a cafe con leche. A few tables away sat the quiet older man that had helped me with my first rock the previous day. He had a 70s-academic look about him, with unkempt grey hair, a dated orange backpack, and hiking boots, I liked him. We nodded to each other when I sat down. It amazed me how our paths had crossed a couple of times and how the Camino was providing such interesting encounters.

When my coffee and smoke were finished, I felt rested and ready to start back. I stood up and loaded my pack on my back with my quads fighting against the idea of moving again. The steep hill had pushed those muscles to a place that I would not normally have used. I am not much of a complainer, but I was starting a tally of the body parts that no longer felt comfortable. When I walked away from the village, I was astonished by how my body even agreed to continue. "Mind over matter," was a phrase I knew well and was most useful at that time as there was nothing else I could do but simply walk.

It was beautiful walking amongst the vineyards with their branches laden in the fall grapes. I plucked and savored as many as I could, they were such a succulent treat in the warm sun. I ventured through tunnels spray painted with words of encouragement from previous pilgrims. I traveled through the various little communities, rewarding me with a sense of accomplishment, only to then find myself on another wilderness path. When I encountered other pilgrims, we would say, *"Buen Camino"* or "Keep going you got this." Even in my solitude, those people and words made me feel not so alone. Harold, my walking stick, not a cane, supported me when I was tired, when the up was steep he pulled, when the down was daunting he kept me from falling. Without him, I would have certainly struggled more.

I nibbled on the last of my nuts from the man in the forest, smiling at the thought of that encounter, and walked the last 10 kilometers of the day in silence. I appreciated the sound of my breath and the gentle thump of my beating heart, I didn't remember ever doing that before. I arrived in the town of Estella just before nightfall. I had walked 27 kilometers that day, realizing walking in double digits was my new normal. I was sincerely exhausted and my body was begging me for something other than walking. Inside my boots, two very painful blisters were forming on the outside of my heels. I approached the first *albergue* in the town, armed with a bottle of medicinal wine for minimal euros, and checked in. Instead of heading to the shower regardless of how much I stank (deodorant doesn't stand a chance on the Camino), I was more interested in the sound of laughter coming from the courtyard. I plunked myself down at a table with another woman who was nose deep in a book. I offered her a glass of my wine as I was unscrewing the cap and she peeked up from her page and nodded yes. It took nothing to share what you had with your tribe and it reminded me of the woman in Zubiri with the anti-inflammatories. Everyone in the courtyard had their boots and socks off, and the display of blisters, swollen feet and bandages was almost comical. I was not alone as I removed mine.

"Crap," I said out loud as I openly laughed at the various blisters adorning my toes. There were new ones over the old ones, along with two sausage shaped ones I had felt on the outside of my heels. My laughter brought inquiring pilgrims, as like myself, they were curious to see the condition of my feet. There was Patrick, from Ireland, covered in Celtic tattoos, Ben from Roncesvalles, Sarah from South Australia, long-legged Alex from Columbia, and, much to my delight, Jaro and Olga. We all spent the evening drinking the wine we had brought and eating Korean food made by a young man from there. It was a festive evening full of advice, laughter, and singing in languages from all over the world. It was the best escape from the demands of the Camino.

The night came to an end when the owner of the *albergue* bantered in angry Spanish for us to all go to bed. As we said our goodnights, the young Korean man who cooked for us all, gave me a beautiful scarf and Jaro handed me a classic Russian playbook. In fits of giggles and drunkenness, we scampered away from the innkeeper and made out way to our respective dorm-style rooms. I sauntered, actually weaved, my way to the communal bathroom and showered quickly, clothes and all, to relieve myself from the dirt of the day. Finding my bunk with the lights out, a little inebriated, was no easy task but I succeeded. I tied my freshly washed clothes around the top bunk to dry, found a plug to charge my phone, and crawled into my cozy purple sleeping bag. Thankful for the wine in my system and the warmth from the night of friendships, I quickly fell into a deep sleep.

HUNGOVER ON THE CAMINO...

The buzz of activity woke me from my first deep sleep in months. It was my luck to be surrounded by diligent pilgrims who had not taken part in the festivities the night before. I reluctantly crawled out my sleeping bag and joined the herd, no sleeping in on the Camino.

It was still dark when I went outside, with only a glimmer of the trusty yellow arrow on a wall to guide me left. My legs were so stiff and my boots felt like blocks of cement, the argument between mind and body commenced. I was questioning whether I'd walked too much the previous day or drank too much. Stumbling into a petrol station just outside of Estella, I grabbed a coffee, muffin, and a bench. The urgency to walk subsided as I had the ingredients to officially wake up. I posted some fun videos and pictures from the nights' party on Facebook and waited for the sun to appear. James, my coffee friend from back home, was quick to respond. His message was that of questions and jealousy regarding the video of me hugging and dancing with Patrick from Ireland. I felt uncomfortable with his reaction, as I did not belong to him. I subdued him gently, but firmly, reminding him of my freedom and how I didn't want to feel attached to anyone, in fact, it was what I was walking away from. I was still recovering from the emotions of Him, and even someone as kind as James could not soften that transition.

When the sun was finally shining, it was time to start the day. I checked my Camino app and saw there was a 12 kilometer void of civilization up ahead. The morning rush of pilgrims had long passed me while I sat on the bench, so alone again I would be. Friends from home asked me if I would be walking with a tour or group of some sort and when I responded with, "nope, on my own," there was always a look of concern and curiosity. I, however, thought nothing of it, living with Him had been a lonely existence.

I heaved on my pack, pinned with drying underwear and socks, and set out, wearing shorts and a tank top, exposing my browning skin in the beautiful fall sun. The arrows led me along gently rolling hills with relatively flat sections and through quiet fields for most of the day. I discovered a rhythm of one foot in front of the other that flowed without much thought. The purple monster had become a part of my body and no longer felt like a separate entity.

Just before the village of Villamayor, I came upon a 13th-century cistern where I knelt grateful to drink the cool fresh water. There were many of these along the Way, created in a time when pilgrims walked in robes and sandals with nothing more than a pigskin satchel—a far cry from my ergonomically correct backpack and hiking boots with padded insoles.

It was just six kilometers before Los Arcos, where I had planned to stay the night, when I came across a food caravan out of nowhere. There had been nothing for some time and the little oasis was a welcomed reprieve. I stood at the counter and waited for my cafe con leche admiring the walls inside that were covered in pictures of the proprietor on the Camino.

"How many times have you walked the Camino?" I asked.

He replied with a chuckle, "I have Camino blood. I have walked it eleven times, no longer on a pilgrimage, I walk for the love of it."

"Wow, that is really amazing," I said.

"It is where I find the most honest, caring, and rare breed of people," he said, pointing to the many smiling photos.

"I could not agree with you more."

I thanked him and when I turned around there was Olga and Jaro coming in off the Way, I was delighted to see them again. We gathered at one of the little tables set up in the dust and discussed the outcome of too much wine and walking, laughing from our bellies.

"We must leave the Camino in Los Arcos," Jaro said to me. "We are disappointed to not be able to continue but our new jobs await."

"It is strange but I will miss you both, though we have known each other for such a short time," I replied.

"That is what the Camino has done for us, friendships found out of nothing more than a dedication to walk," added Olga. "Meeting you Tess has been such a joy for the two of us. We will think of you often on your walk to Santiago and we ask you to carry us in your heart as you go."

I promised I would hugging them both goodbye. As I walked on towards Los Arcos, I was feeling like a true pilgrim becoming ingrained into the Camino and all it had shown me already. It had taken a deeper significance than when it began and I felt stronger than ever.

640 KM

I woke up the next morning very grumpy. What the hell was I doing here? Despite forking out a small fortune for a bloody fancy hotel, I slept horribly on a bed that felt like concrete. When I did briefly doze off, I had weird dreams of people from my past entwined with people who were now in my life. The dreams ran like one of those drawings done on a pad that when flipped through show an animation. There was nothing specific I could remember, only that I saw people I had not thought of in a long time, which was weird. I got dressed quickly, stuffed my pack with no organization. I wanted out of that strange town, which was messing with my mind. I stopped only briefly to grab a cafe con leche to go from a grumbling barista, whose mood reflected the energy of the place. Going under the ornate archway signified my departure from Los Arcos, which immediately lifted the weight of my dark, negative mood. I turned only once to look back from where the sun was shining all around me, Los Arcos remained in the dark, I felt a cold shudder up my spine. I turned towards the Camino and, with Harold swinging in step, started the day. Gone was the question I had woken up with.

That day marked one week on the Camino with 150 kilometers of walking behind me. I passed a sign reading 640 kilometers to Santiago and found it impossible to fathom where I would end up. Knowing there was a destination was part of my motivation, I always did my best when there was an end game. The line of work I was in at the time was interior house painting. I got great satisfaction in walking into a room in the morning and transforming it into a fresh new color by the end of the day. Much like my relationships prior to Him, but he was the first time I didn't want the end to come, even when it had.

I arrived in the town of Sansol after completing seven kilometers and took the opportunity to sit down with a cafe con leche and a smoke. My physical discomfort had become manageable with blister bandages, ankle wraps, a dose of anti-inflammatories, and staying positive. Across from the stoop I was sitting on Dunja was having breakfast with four other people. Puffing with enjoyment of my Marlboro, I swear Dunja was giving me a disapproving look and after our intimate experience with the vultures, she hardly seemed to acknowledge me. Normally I would have felt self-conscious and stubbed out my cigarette, but no longer was I going to allow insecurities to engulf me. I remembered a quote I once heard that I now finally understood, "It is none of your business what other people think of you." Instead, I became an observer of the four people Dunja sat with, my favorite pastime, people watching. I figured they were two separate entities, a mother with her daughter and father with his son. It seemed obvious there was a connection between the daughter and son, a little Camino love affair possibly? The mom, who I heard addressed as Martha, was a vibrant, happy person as she engaged with laughter and smiles in her conversations. Observing the dad, who I overheard addressed as Rodger, was the type of man with an inflated ego. Mostly because I could hear extreme criticism in the way he boasted about his various opinions. After dissecting politics and religion, he moved onto the movie, *The Way*, berating the main female actress. I felt defensive for the character she played as I could relate to her role in the movie. She had been heartbroken, too slim, her face showed years of a tough life, and she smoked the whole Camino, simply trying to do her best. I liked her character and the realistic role she played. Rodger in his deep, loud voice managed to turn her into a scraggly, unattractive stupid woman that he would never give the time of day to. I was certainly eavesdropping on the conversation and had to stop myself from thinking he was directing the words to me, I knew it was a ridiculous thought. That was the kind of overthinking we all do in our heads that doesn't serve us. I went back to remembering the previous quote. I gathered up my things with awareness to remove myself from situations that would encourage those thoughts. I turned to the Camino and plugged in my headphones with AC/DC blasting in my ears, *Highway to Hell*, how appropriate, I laughed.

It was 11 kilometers to the next village when the steep hills began again. The arduous climb was a real leg breaker with a never-ending successions of hills and the wind had become an unpleasant companion. My calves and quads burned and my lower back ached as I admired the sheared grain mountain field and changing colors of the random vineyards. Then the Camino gave me a break, as Martha was walking in stride with me. We first exchanged the normal pleasantries. She was from Rhode Island in the United States and yes, she was walking with her daughter and yes, indeed there was a romance blossoming between her daughter and Rodger's son. The foursome had met a few days ago and the kids had hit it off and were now walking together.

"Rodger," Martha explained, "had tried to 'hit it off' with me but to no avail. He just never stops talking, and what he talks about is so negative so I stride ahead of him constantly. I even told him that when we first met but he doesn't seem to get it."

"I am trying out at being a bit selfish, only getting to know people that lift my spirit, rather than being overly polite," I said. We both agreed we didn't have to engage with every person we meet.

The conversation went to swapping mother stories, Camino stories, and the big life changes that would come when we got home. Martha explained she was contemplating moving to another state to be with her longtime boyfriend. I, on the other hand, told her I had no fucking idea what I was going to do. We both laughed at our opposite dilemmas and Martha was kind enough to assure me that my plans would come together when I was ready. The link we felt for each other in such a short time lifted our spirits. As we walked and chatted, we came across an older man standing on the side of the path. He looked at the two of us with a big toothy smile peeking out from under his snow white beard.

"Are we in Santiago yet?" he asked us.

I couldn't help but break into song, "It's a long way to Tipperary, it's a long way to go." Laughter erupted from all three of us as we stood alongside the man who was quick to introduce himself as Harold. (I smiled thinking of my walking stick)

"This is by far the warmest Camino I have ever walked," he said.

"How many times have you walked Harold?" asked Martha.

"This is my twelfth time, though I have only completed nine. Half of those were with my wife before she passed away fifteen years ago, now I walk alone with her spirit." He shared with his hand over his heart. "I am eighty-five years old now and I know this will be my last Camino, and I need to finish this one."

I thought to myself, if I wanted to walk into Santiago with someone it would have been Harold. Martha and I listened to abbreviated stories of his previous walks, especially enjoying when he spoke of his wife. It was delightful to listen to the love he cherished for her as it radiated through his words, which beautifully expressed what a gem she had been. To hear about such a love gave me a wistful feeling of what I had hoped I would have found at fifty-two. Harold interrupted my thoughts by politely asking if either of us had any extra water, he explained he had run out and was concerned with the distance to the next town. Without a thought, I shared half of what I had left in my bottle, I was honored in doing so for that lovely soul. After the exchange, I was ready to move on, I could feel my ankle getting stiff and it was obvious Martha had more questions for Harold. I bid them both *Buen Camino*, getting a huge Rhode Island hug from Martha, and a warm encouraging handshake from Harold.

Not long after saying goodbye, I summited the mountain climb and was presented visually with what goes up must go down. The descent was as excruciating on my knees as expected, so I positioned myself to walk down doing mini switchbacks to accommodate the discomfort. That plan did wonders for my knees but very little for the blisters on the sides of my heels. Regardless of the soreness, I admired the beauty of nature and the breathtaking views off to the horizon that the elevation allowed.

I noticed ahead an older Spanish man who was obviously struggling a great deal down the narrow rocky path. As I came closer I observed his attire was questionable as far as walking the Camino standards were concerned. He was carrying a heavy winter coat with a makeshift pack on his back and mismatched socks inside his sandals. Curiosity, not judgment, triggered my empathetic heart to stop alongside him and hold out my arm, which he took, giving me the softest smile. Together we conquered the difficult path, reaching the bottom with not a single word spoken. When it leveled out, he turned to me with a broad toothless smile and gave me a fatherly hug. It would be one of my favorite moments on the Camino that I would never forget. I, in turn, silently offered Harold, my walking stick to the man but he bowed his head elegantly and whispered, "No *gracias*." As I walked ahead, I smiled at the interaction with the angelic soul and acknowledged in myself how good it felt to care for others. A wonderful warmth encased my repairing heart.

Arriving in Viana, a small town dating back to the 14th century, I decided to call it a day. It was more than nine kilometers to the city of Logrono, where I had intended to finish the day, but the charm of Viana encouraged me to stay. I walked the narrow streets, passing stone houses, small wooden cottages, and many smiling, cheerful residents, it was much different than my arrival into Los Arcos the night before. I found my way into the town square, a destination that had become the highlight after the day's walk. I really loved that each place had a community square and it was there I would embrace the energy of not only other pilgrims but the locals as well. In Viana, being a smaller town, the nucleus center was much more intimate than the others I had stopped at. My gaze caught the professor sitting alone against a wall in front of the fountain, I crossed the square and sat a few feet away from him. Not being a man of many words, we nodded at each other in recognition as I sat down. I could not help wondering what his story was observing him with his ancient orange pack and content serene look on his face. He was a peculiar sort of man in an endearing way. My voice broke the silence.

"Have you found a place to stay yet," I asked hoping to discover more about him.

"Yes," his voice spoke slowly in almost a whisper, "beside the church over there." He pointed.

"Is there room for me do you think?"

"There is, but you might do best to go to the large municipal *albergue*, I have noticed my dorm room is full of men."

"Okay, thank you," I replied smiling.

We sat there a while longer before he stood, nodded to me and, with his long ambling strides and fingers hooked in his pack ropes, disappeared. A man of few words noticed me, solo on the Camino, and cared enough to give some good advice, not so much for safety but more for my comfort. I had felt appreciation for the man who cared about my welfare.

On my trek to find the *albergue*, I passed a small storefront with a sign offering pilgrim blessings. I entered the store and was greeted by a big bear of a man who enthusiastically reached out and gave me a hug.

"Welcome dear girl, come sit down. Would you like some tea," he asked.

"Yes please, that would be nice, thank you," I replied.

I removed my pack and plunked down on the unoccupied overstuffed purple couch. Across were two big eclectic armchairs and a beat-up coffee table laden with candles, books, and cookies. It felt good to sit on something so comfortable, in a home like atmosphere.

"What exactly is this place?" I asked the man as he went about steeping the promised tea.

"It is your place for the moment you are here. A place to rest your tired sore body as you prepare for the mindful stage of the Camino to begin. Please let me introduce myself, my name is Terry. My wife Lauren, who is out right now, and I walked the Camino only to return to this place to take it over from the previous pilgrims. That was two years ago."

"Where was home before coming to Viana," I asked.

"A small town in Ontario, Canada," he replied.

"That is incredible, I am from Vancouver, Canada." I smiled at Terry. "My name is Tess."

"What a pleasure it is to have a fellow Canadian come in, so nice to meet you, Tess."

"I am glad I stopped, it's funny how you get a feeling to just do something." Terry handed me the tea and sat down in one of the armchairs across from me.

"Life is about moments, and sometimes a moment becomes a life. That is how Lauren and I feel about what we are doing here. We offer refuge for you to come to terms with the physical demands of the Camino before you embark on the next stage of mindfulness."

"I have heard of the stages and am aware of what I've put my body through to get this far. I have been so overwhelmed at moments that I could cry," I shared with the kind man.

"Rest assured that is a normal feeling, along with the loneliness that can be felt. It is impossible for anyone outside of the Camino to understand what you are going through. That is why you are sitting here and verbalizing as you need. However," Terry added, "I will not let you wallow, instead I would like to offer you a blessing of additional strength to complete this journey."

He then reached behind the chair into the bookcase and withdrew a well-read leather bound book. Opening it he leafed through the pages until he found what he was looking for.

"This book has the blessings in all the world's languages, it has been here in this retreat for a very long time. Sometimes, finding and properly pronouncing the dialect for some is the hardest part of my day," he laughed. "I am happy to do English when I can." I grinned and bowed my head as he began to read from the ancient text. He spoke with a voice that caressed my spirit and infused me with a feeling of assurance. The words were exactly what I needed to hear and, when he finished, I reached out and laid my hand on his arm in gratitude. I reluctantly stood, leaving the comfort of the cozy couch and heaved my pack back on. Briefly, my thoughts went back to Los Arcos, the town of darkness, as I compared it to the place I was in. I understood better the experience allowed me to see the yin and yang of the Camino. For without balance there is no awareness, which is all a part of becoming a whole. I was part of that. Another pilgrim came in the door, so I quietly slipped out to allow her to embrace the sanctuary of that place.

I found the *albergue*, paid five euros, and was handed a disposable bed bug sheet (oh!). I climbed the stairs slowly, there were lots, and entered my first large dorm style sleeping arrangement. Rows and rows of bunk beds lined the walls of the drafty room. I selected a lower bunk close to the door, grabbed my toiletries and headed to the communal showers. Thankfully I had it all to myself, as I commenced the process of washing hair, skin and clothes. Afterward, wearing the one sundress I brought with me, I ventured out for some nourishment. Walking through the quaint streets, I came across Neil and Tatiana, a sweet couple I had met only briefly at dinner in Puente Reina. They were sitting at a table outside a small restaurant, and recognizing me, enthusiastically waved me over to join them. Together we drank wine, ate wonderful food, played cards, and shared a lot of stories. It was a wonderful evening. When my exhaustion took over, I excused myself to pay my bill but instead got a big hug from Neil and was told to go to bed. A gift from the Camino, again.

Back at the *albergue*, it was obvious by the darkness that everyone had gone to bed. I crawled straight into my sleeping bag, my sundress serving as a nightgown and drifted off.

Sometime in the night, I was abruptly awoken from a deep sleep with a horrific weight on my chest, a stranger was aggressively groping both my breasts. I sat up too quickly in the shock and hit my head on the lower bunk above me. I did my best to stifle a scream, though it was enough to send him scampering off into the darkness of the still sleeping dorm. I was shaken to my core as I tried to come to terms with what had just happened. Not knowing what else to do but run, I hastily gathered my belongings and scurried down the stairs. Unlocking the massive wood doors and dragging my sleeping bag and pack behind me, I went out into the night air. Wandering aimlessly in the dark I was feeling a level of low I could not imagine existed. All I wanted was to go home but was quick to realize that home was no longer there, I had nowhere to feel safe. The coldness enveloped my bare skin in the thin sundress and I shivered against the air while dragging the purple monster along the ground. Following a bumpy stone path I found a grassy area and slumped down on the green carpet and sobbed. How could this have happened to me? Had I not already had enough? It brought back memories of my father, memories I struggled to steer clear from. Emotionally overloaded, with no ration of logical thinking, I crawled inside my sleeping bag right where I was. My best bet then was maybe a divine being would watch over me while I slept under the stars.

DIGGING DEEP

I opened my eyes the following morning to the smell of dew under my nose and the most brilliant sunrise I had ever witnessed. Stumbling my stiff body out of my bag, I embraced the amazing beauty that was right before me. In the darkness and confusion of the previous night, I had placed myself on top of a decaying turret overlooking the entire landscape surrounding Viana. Beneath the rising sun the city of Logrono was about 10 kilometers away. I had a choice to either let what happened the night before go or hang onto it and infringe on the journey I was on. Thanks to the sun, my ever trusting connection to life, I packed up my gear and headed into town. Sometimes, walking away is the best lesson, so I did.

Unfortunately for me with the extremely early morning start, I quickly discovered there was not a drop of coffee to be found. For an obsessive morning coffee person, it made the stretch to Logrono even longer. I walked the long, boring flat path along the interstate with only a couple of cigarettes. My ankle was angry and protesting again. I tried adjusting my step by putting more weight on my left foot, which only resulted in infuriating the nasty large blister on the outside of my left heel.

"Shit, shit, shit," I muttered to myself.

When I arrived in Logrono I felt relief, mainly because I knew there would be coffee in my future. I was quick to notice it was a large industrial city that attacked my senses. I followed the painted arrows on the sides of buildings and light posts that led down the narrowing streets into the original part of the city. There the bustle subsided, like stepping into another world, with small outdoor cafes strewn around. I saw Neil and Tatiana and went to join them at their table, ordering a cafe con leche from the server when he approached. I was grateful to have the opportunity to thank them again for dinner. I did not share what happened the night before. I felt no need for the sympathy that would follow, rather I chose to enjoy the morning banter we were quick to exchange instead. Neil offered I walk with him and Tatiana for the day, but I declined. I knew from the 10 kilometers that my pace was going to be slower that day than theirs and did not want to infringe on them. In the world of Camino life, pilgrims will band together when needed and leave alone when they must, simple respect.

After my third cafe con leche, I continued walking through the city admiring the beautiful buildings of various masonry. One of them, a church, Iglesias de Santiago, held my attention as I was intrigued to go inside. I had done some reading about the churches on the Camino and that one I clearly remembered. It was constructed in the 16th century, built on the foundation of an even older church that had been destroyed by a fire. Crossing the threshold, I was quick to observe the interior resonated a Renaissance/Baroque style with an impressive ornate altar. I knelt in a pew, instinctual from my Catholic school days, and asked whoever was listening to help me forgive the men in my life that harmed me. I used that opportunity to embrace the strength I knew I had, to understand more about what it meant to let go of something that did not serve my higher good. Satisfied with my request, and feeling lighter, I left the church and began the second half of the day's walk. Leaving the city was far more pleasant than when I arrived, as I walked on a path through a beautiful lush park that banked alongside a large river. The only unpleasantry was the insane burning sensation escalating inside my left boot. Unable to step away from the pain any longer, I shrugged off my pack and sat down on a bench alongside the river to have a closer inspection. I removed the boot and stripped off my sock and I was aghast at the discovery. The somewhat small blister on the outside of my left heel that started in Los Arcos had grown to the shape and size of a pickle.

"Crap, bloody crap."

I dug in my pack and pulled out the unsophisticated first aid kit I had assembled back home and began the necessary treatment of the offending blister. I sterilized the safety pin with my lighter and poked six holes into the pressured skin, which encouraged the nasty fluids to escape. The understanding and sympathetic looks I got from people passing by gave me encouragement as the burn slightly subsided from the release. I had no previous experience dealing with a blister that size, so I went through a logical process of draining, applied antibiotic cream, bandaged, and wrapped with Elastoplast in hopes my work would stay in place. When the mission was accomplished, I put my boot, painfully, back on, hoisted the purple monster into position, and, with Harold by my side, took a step only to growl from the deepest part of me from the pain.

"This is bad," I cried out. "Really bad."

Given no other choice but to accept my fate, I limped along with my head down, focusing on the 13 kilometers to the next town. I could not even respond to the "*Buen Caminos*" offered, I was zeroed in on walking, to simply keep going. Somehow, Neil and Tatiana appeared at my side and offered their assistance. I could hardly express myself but assured them I would be okay. I was a pretty independent sort of person and my mind was in a place of solitude I needed to reach my destination. Had there been a village along that stretch, I would have stopped, but there was nothing, only the never-ending hill climb that created a bonfire in my boot. I was on day nine of the Camino and it was by far the toughest. Harold was no longer my walking stick, he was my cane with every step. The loneliness I felt overcame me like a tsunami and I had thoughts of just laying down and dying, I was in an altered state that no one could understand because I certainly didn't. Every step was painful and my body was trying to shift itself to handle it, which only resulted in new muscles tearing apart. I pushed on until I could no more and crumpled at the side of the path. I ripped off my boot and sock to see the carefully wrapped wound and gasped in astonishment. A new blister had formed on top of the one I had recently drained, the same size and displayed colors of red, blue, and black. The strain of the pain with the visual became too much for me to handle, something had to give, the tears needed to flow. It was then I looked up to the sky with red-rimmed eyes and called out to my Mom.

"Please help me, Mom, I am scared."

I dropped my head in momentary defeat and reached for my first aid kit to repeat the same process I had done only hours before. When finished, I angrily pulled my sock and boot back on and stood up. There, directly in front of me, was a big blue butterfly sitting on the path—my Mom. I smiled through my tears and dust and somehow, out of sheer willpower, tackled the next three kilometers to Naravette.

What should have taken me forty minutes, took two hours at my pace. I staggered into town, passing pilgrims who were showered and on their third glass of something cold. I went in search of a decent hotel determined to reward my body with the best care I could give it. Lying on the queen size bed with my feet in the air, for they could not even handle the pressure of the bed, I waited for the pain to subside. As I waited, exhaustion overtook me, and I dozed off in a fetal position.

Several hours later, I opened my eye and hoped the anti-inflammatories had worked. I stood up to test out the weight bearing factor. The pain was tolerable and I was in need of food so I scoped out what was around me to get me out the door. There was no way I could put my boot on, so I mutilated my left Sketcher, cutting away the back of it so it would not rub on my heel. I slipped into it like a slipper and used some Elastoplast to secure it to my foot. With Harold by my side, and handrails, I found my way downstairs to a cafe. What luck, sitting there was Neil and Tatiana. Neil saw me first and jumped up with a caring smile to take my arm and lead me to Tatiana and her warm hug.

"You are here, we are so happy to see you. We have been worried about you," she said.

"Aw thanks," I replied as I sat down. "That was the hardest day I have ever experienced physically, I think giving birth was easier." We all had a laugh.

"When we passed you today, your face gave away how much pain you were in." Neil said smiling at his wife, "We both agreed you were one of those tough gals that would find your way, and here you are!"

"It amazed me how the mind can dominate the body, not sure how logical it was out there," I replied, lifting a cold glass of wine to my parched lips, "But look at me now." The three of us toasted my accomplishment. There was no need to recount the day, it was clear to us some stories are better left unsaid. Tatiana handed me a simple woven bracelet with a Camino shell on it, to always remind me of the day I had overcome with strength. We all ordered and devoured a hearty meal and finished a bottle of wine, laughing the entire time. After the plates had been cleared I craved the comfort of a cigarette so much, I had run out the day before and just wasn't ready to quit. I stood up unsteadily to go find some, but Neil also stood up and gently pushed me back into my chair.

"No way kid," he said, and, even though he wasn't a smoke, he trotted off to buy me a pack before I could protest. The kindness and understanding without judgment allowed an amazing feeling. Earlier I had mentioned I had lost my sunglasses (likely in my hurry out the door the previous night) and, without a thought, Neil had taken his own off the top of his head and handed them to me. The Camino was showing me to accept kindness without feeling defeated.

"Neil is a remarkable man," I said to Tatiana while we were alone.

"Yes he is, we have found a love that we know without a doubt and we are grateful to have. The Camino is taking us to a deeper understanding of unconditional love."

"I am so happy for you both. I hope to experience that one day."

"Tess you radiate an energy of unbinding love that both Neil and I felt that night in Puente Reina. Though we did not talk much, we observed how you projected care and warmth like the sun as we watched you interact with others. They sat in awe of what you were giving them without you even knowing it. When we saw you earlier today, we wanted to scoop you up in our arms to help such a beautiful person, but we also saw and respected the fire of determination in your eyes." Tatiana explained as she reached for my hand. "True love is out there for you too, I think you might be giving up, but please do not. Who and what you need is a very special person and that person will embrace all you are and love you back unconditionally. I know this without a doubt to be true."

Neil returned then with my beloved Marlboros, smiling like a proud hunter back with his kill. I gratefully inhaled the pleasant smoke and felt peaceful from the words that Tatiana shared with me.

Later, alone in my room, I messaged a quick update to my kids as well as responded to James who had left several messages wondering how I was. I replied honestly about my condition, with a picture of the detrimental blister, and he was quick to reply that maybe I should slow down and take a day off. He challenged me with a question of why I was driving myself so hard and reminded me it was not a marathon. With my feet hanging off the bed, I drifted into sleep wondering for myself why was I pushing so hard?

SLOW DOWN

The next day I found myself in a dilemma. Should I shorten the day with a seven kilometer walk to Ventosa or take the day off? I evaluated my ankle and legs and realized they did not feel too bad but the concern for the blister was very real, there was no way I was going to get my boot on. Looking out the window, the weather was overcast and I could feel the cold against the glass. Unable to make a decision I resigned and crawled back under the covers.

Two hours passed before I awoke from bizarre dreams similar to the ones in Los Arcos, only this time the movies seemed longer and only of people I was no longer in contact with. I was curious why I was dreaming like that, possibly I needed to relive experiences of those who had passed through my life. Given I was on a journey of a lifetime, and I had opened the recesses of my mind, which were resulting in the weird dreams, I felt a bit unnerved.

I gazed out the window and my spirits lifted as the dreary morning had given way to blue sky. The decision came to take the day off, and with that, I went in search of a pharmacy to resupply my depleting first aid kit. What I found was better than that, a walk-in clinic. In my passible Spanish, a picture on my phone, and hand gestures, the receptionist pointed me over to a bench to wait. It was not long before a nurse approached and guided me down the hall to an exam room. I sat on the table and she removed my modified Sketcher and bandages as gently as possible, explaining I had not been draining the fluid-filled blister properly, which was why it was so aggravated. She proceeded to expertly lance through the layers of skin, which provided drainage, then cleaned the wound, applied ointment, and wrapped my foot. I thanked the nurse and left the room more comfortable than I had come in. I stopped at the front desk to pay for the services and was waved away wishing me a *Buen Camino*. I felt privileged, the Camino again provided.

I made my way back to the hotel with another valuable lesson, one that I would practice not only out there but also in my hectic lifestyle when I returned home. When I started the Camino, it was more of a mission to accomplish it, to get it done. I lived my life that way most of the time, which seemed to result in burning out and/or losing interest in the tasks at hand. That was not an option when I was trying to walk across Spain, I needed to gear down and take each step mindfully. It showed me beauty, kindness, love, and also affirming what a courageous, strong woman I was. Consciously, I pressed a mental rest button and asked the universe to help me carry on by slowing down.

CARLOTA AND JORGE

"Are you kidding me?" I said to the mirror the next morning. I could hardly believe what I was looking at, a swollen, red, weepy eye. It was obvious I had shared the bed with a nasty spider. I grabbed a washcloth, wet it down with cold water, and applied the compress to my eye, thinking what else could go wrong? I couldn't help but giggle at that point, which broke me up into laughing so hard, I had to reach for the sink to keep my balance. "Okay Universe or Camino Gods, enough is enough, I will walk today!" The one-eyed morning to get out the door started: got dressed, wiped eye, brushed teeth, wiped eye, packed the purple monster, wiped eye. With the modified sketchers taped to my feet, the impossible to wear boots hanging from my pack, and cane in hand, I looked at the full-length mirror and smirked at the vision of the poster girl of the Camino.

After I checked out, and paid my extravagant bill, I sat in the small lobby for a minute to adjust the redesigned shoes, when who should pass by but Carlota and Jorge! It was an instant rekindling of the connection we felt when we met and talked in Puenta Reina days before.

"Tess, what has happened to you?" asked Jorge, stifling a laugh at my appearance. Carlota, smiling, was quick to poke him in the ribs in jest.

"Seriously, are you okay?' she asked sitting down beside me.

"A little bit of a tough go the other day. I pushed too hard and paid the price for it. I took a needed rest day yesterday, as a couple of blisters got the upper hand."

"And your eye?"

"Oh, you know random spider bite, taking this whole thing in stride," I replied with a feeble smile.

Jorge reached out a hand to me, "Come on, we will walk together and Carlota and I will show you the 'other' way," he winked at her. The smiles on their faces gave me no other choice but to take Jorge's hand and walk out the door with my new friends. I could feel something very magical about those two, very magical indeed.

BRAVERY

There are three types of pilgrims found on the Camino. Those who walk in large groups, either on a tour or with friends, those who walk in pairs, like Carlota and Jorge, and myself, solo. Sharing the walk that day with my new friends was comforting as we ambled along at my pace. Relearning to walk slower, or as Jorge put it, "the other way," was very beneficial for me. We spent the time easily and openly sharing personal stories and discussing how it is not often we meet people who we instantly connect with, but the three of us certainly had. I learned the two of them had only recently become a couple and walking the Camino was the perfect way to get to know each other. They too had been in disappointing relationships and the Camino, which was so much like life, provided the opportunity to experience each other.

We climbed a steady hill out of Navarrete, which proved taxing in the heat of the sun. As time passed, occasionally Carlota and Jorge would be ahead of me holding hands and it would make me smile to see the blossoming of two people coming together. It was easy with them as there were no expectations. Sometimes we were side by side and sometimes I would fall behind to be alone. I stayed focused on my steps with hopes of not aggravating the blister or the swelling of my eye. I questioned if I was enjoying myself or not, as all I could feel was discomfort, but I believed I would break through soon and start appreciating the surrounding beauty and experience of it all again. I was entering into the mindful stage of the Camino and I was ready to embrace it.

Jorge came alongside me at my sluggish stride.
"Tess, how does it go?"
I pointed down to my pathetic makeshift left shoe that was unraveling

bit by bit.

"Well, honestly I think my workmanship is coming apart."

We both looked down and saw the tape holding the shoe on was wearing thin and keeping it on was becoming increasingly difficult.

"I might have a bit of a problem very soon," I laughed. "That is the last of my tape, maybe I should walk in my socks?"

Carlota backtracked to join the two of us, and Jorge spoke to her in rapid Spanish that I could not understand. When they finished, Jorge took off his pack and reached inside, producing a pair of running shoes.

"Let's try these instead of your socks," he teased, offering them to me.

"Come on Tess, these will be better for the Way," Carlota added with a big smile. I hesitantly took the offer of the shoes from Jorge. When offered a gift from someone with so few supplies it was hard to accept, but it did come from a place of genuine compassion. I slipped into the comfortable shoes and felt the much needed relief. I stood up to test out the new footwear and succeeded in doing a little happy dance for my friends to demonstrate. Then Carlota and I took the hacked up Sketchers and set them on a stone wall with flowers tucked in them. On the Camino, you only carry what you need and those shoes had served their purpose as far as they could. I found it strange to walk away from them but it was what I needed to learn: leave behind what you no longer need. It was moments like that when you learn to live again. Feeling so good over the borrowed pair of shoes, we walked on as Carlota told me about the amazing beauty that lay ahead: vast mountains, endless vineyards, and unique villages. She shared she had walked the Camino once before. I detected a hint of sadness in her eyes but she offered no explanation.

We arrived late in the afternoon to Najera, a beautiful town with a river running through it and welcoming cafes dotting its bank. There were people strolling about, bicyclists, children running around, and dogs scampering everywhere. They all looked so happy, it was a joy to be there. Carlota was able to phone ahead to secure us lodging, something I could never have entertained without a phone plan or fluent Spanish. The two of them escorted me through the ancient architectural town to my booked lodging. When we arrived, I looked at the entrance with a bit of a shock as it was an upgrade to what I had previously experienced. Carlota observed my reaction, "Tess you need a proper place to rest and heal."

"I think you are spot on with that idea." I smiled in anticipation of the comfort I would be in that night. They then told me the two of them had booked a romantic little place not far away. We hugged each other farewell, Jorge assuring we would see each other again as he pointed to his shoes on my feet.

"Oh no! Let me give them back to you."
"I am not worried about my shoes, little sister, I only worry about you," he replied.
"Little sister... I like that big brother. Thank you."

After we said goodnight, I checked in, received another Camino stamp for my passport, and made my way up the stairs to my room. I was sad to see Jorge and Carlota go—the first people I felt a connection to in a long time—but believed, in the Camino way, I would see them again. Hungry and thirsty, I dropped my pack on the bed of my private cozy room and noticed a bottle of wine on the bedside table. On the bottle was a note attached and in English it read, "You are strong." I was not sure if Carlota had arranged that or it was customary in that place, either way, it was thoughtful. It would be a treat when I returned from filling my belly and replenishing my now limited blister dressings. Back into the core of the town, I ventured.

The streets of Najera were winding with no parallel pattern and invited a foreigner to get lost. I have always had an innate ability to remember focal points in strange places so as not to go astray so, with confidence, I went in search of a pharmacy. I passed open bars and cafes where the locals laughed, sang and conversed, it warmed me to witness the liveliness. Finding what I needed nestled between two dwellings, I walked in and approached the smiling pharmacist. It was immediately obvious we had a definite language barrier so I used sign language and photos on my phone to describe what I needed. The pharmacist spun on his heel with a paper bag in hand and began pulling various things off the shelf behind him. Once he was satisfied, he handed the bag over to me at a minimal cost. I wasn't sure what I had bought but felt confident I had all I would need. The next stop would be the food my body was hankering for.

I strolled along the river, passing outdoor cafes until I found one with an inviting table. Devouring a small flatbread pizza and a couple of inexpensive glasses of *vino blanco*, I was feeling content. Across from me was a table full of Camino pilgrims enjoying a meal together. There were so many different accents in the air and I enjoyed watching their enthusiasm as they would finally understand each other. Finished with my last sip of wine, I retraced my steps back to my temporary home for the night. Once settled in my room, I sent a brief message to my kids, though slightly edited, about the last few days of adventure, and a note to James that I had indeed taken a day off. It felt good to touch base with them. The next step was to pop the cork on the bottle of wine left in my room and take several gulps in preparation for the required bandage change. I emptied the brown bag from the pharmacy onto the bed and quickly calculated I had enough ammunition to perform minor surgery, so I did. With years of experience dealing with horses and their injuries, it was like the pharmacist knew what I needed to do. Unwrapping my foot from the confines of the bandage, I was grossed out by what I saw. Another gulp of wine was needed. I began the painful process of removing the dead, moist skin surrounding the blister that had grown larger than life. By exposing the wound, I saw it was two skin layers deep, red and very angry. I gently rinsed it with saline, took another gulp of wine, applied generous amounts of ointment, and placed strips of Compeed plasters over my work. Next, I cut strips of white tape and then carefully arranged gauze squares over the wound and secured them in place with the tape. As far as I was concerned, I was a fucking hero. Satisfied with my work, I enjoyed a tasty Marlboro on the small balcony with a final rewarding glass of wine. A wonderful numbness came over me and I took this as a sign it was a good time for bed.

DOG POO IS GOOD

Sitting in the sun the following morning in Najera, while enjoying my cafe con leche, I started doodling on a napkin. I created a beautiful twisting feather with the tips of it dissolving into birds flying off. I was careful to find a secure place in my pack to keep it, I would have it tattooed on my arm underneath the "Trust the Journey" tattoo already there. I felt the power of that doodle and it would commemorate the journey, a sense of my wings evolving to freedom from within.

The walk that day would take me to Santo Domingo, 23 kilometers and, according to my app, several serious hill climbs away. Leaving the comforting town of Najera began with exactly that, a steady uphill climb. With the minor surgery the previous night, I was once again attempting to wear the very expensive boots, with Jorge's runners hanging from my pack as back up. There was nothing smooth about wearing them again, but I felt determined to try because that is why I had bought the bloody things! The only comfort they provided was support for my angry ankle and sore right hip that had been compensating for the irregular gait caused by the blister. I was aware then that there was nothing prior to the Camino that could have prepared me for all I had experienced to date. The daily grind of steep climbs, long ascents, unruly terrain, and unforgiving hard ground had become my life, along with body discomfort and mental strain. I had the opportunity to speak with a few people who had done months of training for the pilgrimage and they too were suffering from the lash, I was not alone. No training or lack of could make the perfect Camino, that I knew. It really resembled life in so many ways. I had always had a driven mind and, when I set my sights, I would push myself hard to reach that goal. Failure had always been a hard pill for me to swallow, thanks to a mother who put high expectations on me to succeed, but that driven mind was what would

59

get me to Santiago. Regardless of the physical strain, I chose that day to embrace the heavenly surroundings of fall-colored vineyards, gold encased grain fields that rolled away like ocean waves, ancient structures dating back to the 9th century, and the aroma of a freshly made cafe con leche. The happiness I felt that day of just being there was surging through my spirit, awakening me to a whole other level.

After a few hours had passed I saw ahead the already familiar silhouette of Carlota and Jorge. I joyously whistled out to them and they both turned around and waved with big smiles. Having caught up with them, we decided to take a break under a huge tree alongside the path. Carlota shared with me the history of the walk into Santo Domingo with its rich traditions of St. James, the patron saint of the pilgrim's way. Long ago, the nuns of Santo Domingo dotted on this stretch, blessing the pilgrims and offering fruit as they passed. Carlota explained the intent was to aid in the long climb before the descent into the town that required strength of mind. I felt so privileged to be in the company of Carlota who could share the amazing stories of the Camino with me.

Once back on our path, Jorge and Carlota walked together ahead of me, though we remained close enough to share our encouraging energy. I enjoyed the simple act of watching them walk together, observing their love blossoming as they laughed face to face, held hands, and supported each other. I was bearing witness to two incredible energies falling in love, something I needed to see to begin a renewal of my mind of what true love should be. Having crested the last large climb with a great deal of determination, we could see in the distance the tall cathedral steeple stretching into the sky like a beacon of welcome. I immediately felt the joy and relief that, after seven hours, the destination was near. With the last three days of darkness behind me, that day had been better and the sight of Santo Domingo had brought the light back into my desire. I hustled my stride to catch up with Carlota and Jorge and together we walked into town but not before I managed to step in freshly deposited dog poo. I hopped around on one foot appalled at myself, but Carlota was quick to tell me I would have good luck from thereon. Scraping my boot along the grass we all broke into belly laughs, the kind of laugh that brings your vibrations up. What else could we do, it was a sight to behold.

Once inside the town, I bid them goodnight and went in search of the nunnery I planned for my night's stay. Entering the old convent, I could not help but feel the holiness of that place and the respect for which it stood. I approached an elderly nun sharing a warm smile where I signed in, had my Camino passport stamped, and was handed an ancient brass key and directed to the antiquated elevator that would take me up to my room. It was a room that could only be expected of a nun to need. It consisted of a single sagging cot, a straight back wood chair, a small dresser, and a simple crucifix hanging above the bed. I had seen rooms like that when I was a child attending Catholic school, so it felt strangely familiar. I gently pulled off my boots and slipped into Jorge's runners and set out to explore the beautiful town.

There was no modern area to Santo Domingo, it retained its ancient appearance in every way. On that night, the streets were alive with locals singing and playing instruments as they paraded along. As I poked about the serpentine of streets, I came upon a small shoe store and had a brilliant idea, sandals! I was confident that would be the solution to anything rubbing against the blister and I thought, if it worked for, then it would work for me. The proprietor was most helpful as he not only understood my broken Spanish but also selected exactly what I needed. While I tried them on and tested them out, who would I see pass by the window but Eva, the fellow pilgrim from Holland. I dodged out the store in the unpaid sandals, called out to her, and we embraced as long lost friends. We quickly exchanged stories of the last few days and I learned Eva too had gone through some dark days. We shared our private moments of questioning whether we could carry on, knowing we could but still had to question, and in sharing, we felt less alone in those thoughts. The simple gift of reconnecting with each other was truly just that, we fed each other in that brief time a new energy. With hugs goodbye and "see you another time"s, I went back inside and paid for my new sandals, adding a neck buff to keep me warm in the coming days of the fall coldness.

Leaving the store, I could comfortably go explore in my, aptly named, Jesus sandals and take in the sites. As with every town I spent the night in, I went in search of the cathedral. The Cathedral of Santo Domingo de la Calzada was named in dedication to Dominic de la Calzada who built a bridge, hospital, and *albergue* for pilgrims traveling the Way. It was he who began construction of the town's cathedral in 784 before he died. Within the structure was also the historical miracle of the "hanged man," a pilgrim wrongly accused of theft. A devout German couple was making the pilgrimage with their son when a young local girl took a liking to the boy but he, being a good Christian, did not respond to her advances. Hurt, she put a silver cup in his satchel and accused him of theft. The town believed the accusation and the boy was hanged. The brokenhearted couple continued to Santiago but on the way back they stopped to visit their son's body and to their surprise, he was still alive apparently hanging from the rope. The couple rushed off to the sheriff and demanded their sons' release.

The sheriff had just sat down to a chicken dinner and laughed at the couple stating, "That boy is no more alive than these chickens on my plate." Apparently the cooked birds jumped up off the plate and started strutting around the dinner table. The sheriff abruptly stood up and went to release the young German boy, who was immediately pardoned and allowed to go home. The witnesses of his successful appeal are represented to date by a pair of chickens kept at all times in the choir loft of the Cathedral. I explored the spectacular structure and found the caged chickens, which made the story more intriguing. I was amazed as I caressed the stone walls, to think about how many people had passed through that place over the centuries.

Once I had taken in all I could, I found a cafe with the customary outside tables and enjoyed a light dinner. Sadly, at the table near me, I overheard a young couple in a heated argument, he was on a marathon Camino and she wanted to slow down. In anger, he got up and stormed away. I watched as the girl dropped her head in disbelief and sadness, my heart ached for her. It appeared their sights on how to "do" the Camino had become different. The Camino defines who you are from within rather than who you think you should be. It guides you through greatness you never thought possible with every humbled step taken. But, you must be humbled away from your ego to embrace that feeling. She was there, he was not.

UNCONDITIONAL

Waking up the following morning in the simplicity of the convent room, I knew the day was going to be a good one. No longer did I feel the rush to get up and go. Having learned "the other way" from Jorge, I laid in bed for a while longer. I remembered a dream I'd had that night and took pleasure in reliving it for as long as I could.

It had been two years since my mother had passed away. Since her death, I always thought I would feel her formidable presence around me. We had a shared belief that after death the soul could still connect with those on earth, so I had been in constant expectation. Though with time, I began to give up hope. I experienced a psychic reading the previous spring in Sedona, Arizona and was told my mom was at peace and dancing on the iridescence of a lake, so I figured she had moved on. Historically, my Mom and I had a turbulent and emotional relationship. Though the love was there, it was constantly challenged. Sadly she left this earthly life angry with me and I had been wrestling with that ever since her death. The dream I had that night finally brought me the peace I needed to let go.

I was sitting at a cafe on the Camino and my Mom came from behind and laid a hand on my shoulder. I could physically feel the warmth of her touch in my dream, and the love. She then came around and stared with her beautiful sea blue eyes into mine and said, "Hi." I saw at that moment my mother's face was beautiful and radiant, and I knew then that wherever she was, she was at last happy. For the first time, I saw her more than just as my mother, but also the woman who had to struggle a great deal throughout her life. To finally see her happy, really happy, was the gift I needed to relieve myself of the guilt of being such a tough daughter. I understood then that she had made me that way so that I would have the skills to endure the conflicts in life and become more than just a survivor. It made sense why she had waited to come to me until I was open, vulnerable, and on a journey of awareness. It was such a simple dream but it was a great reward. She had always been with me: in dance, in music, in butterflies, and in my soul, I had just been too closedminded in my guilt to acknowledge it. It was a blissful way to start my day.

I changed the dressing on the still very ugly wound and tried to put my boots on again, but it was far too painful. So socks and sandals it would be and I giggled at the serious fashion faux pas. I learned quickly on my journey that there was no place for vanity, only comfort. I remembered an expression from my horse training career, "No hoof, no horse," laughing to myself I changed it to, "No sandals, no pilgrim." I thought of how pilgrims centuries ago would too have walked in the sandals and the symbolism could not be ignored.

After seven kilometers in my fairly comfortable new footwear, I stopped in a small village for breakfast. Beside me sat the purple monster with the offending boots draped across the top. I was no longer as fond of them as I had been when I bought them. Their added weight on my back did nothing to soften the relationship I was trying to resurrect. A familiar looking couple stopped by my table to ask how I was.

"Do you remember us from Zubiri?" the smiling cherub of a woman asked. My mind flashed instantly to that early morning hobbling out of the hostel and I smiled.

"I certainly do, you gave me those fantastic pills for my ankle."

"Yes, that's us." They both smiled proudly. "How is that ankle doing?"

"Well, as long as I keep walking, it will simply have to function. The pills really do help. I cannot thank you enough for your kindness.

"Good for you," the man chirped in, "What doesn't lay you down, lets you walk on."

"Yup, now just blisters to contend with and I will be skipping all the way to Santiago." The woman reached into her pocket and pulled out a Compeed roll-on stick for blisters and handed it to me.

"Here take this we have tons."

"For real? You are so very kind, thank you again," I smiled.

"*Buen Camino* little Canadian," they said as they walked away together. I would never get over the kindness on the Camino, being surrounded by like-minded people was pure bliss. I thought about all the material tokens I had been given already, the scarf from Korea, the Russian playbook from Poland, vapor rub from Holland, anti-inflammatories from Australia, a bracelet from the States, and borrowed runners from Spain. How appreciative a person can feel for such simple endearing gifts, each served the purpose to make a moment better.

The walk after breakfast filled me with a peaceful determination to make it to the town of Belorado, 18 kilometers further. I climbed up the Orca Mountains and it was a vista to admire and embrace. Ascending up to each individual crest, I drifted through the woodlands in stellar proportions. It was impossible for me to tire of the landscape as I passed through each stage, sometimes changing dramatically in some places. Being very appreciative of history, not so much in books but more in how long something had stood the test of time, whether it be a cathedral, a monument, a bridge or the mountains, I felt a connection to it all.

I came into a very small village after descending the mountains and, with my new slow down thinking, was in the mood for a cafe con leche and a Marlboro. I came upon a quaint home with a table and chairs and a beautiful older couple outside gazing into the sun. I paused in my step to appreciate the beauty as well and then turned to the couple smiling.

"Cafe con leche?" I asked.

The woman first looked at her husband then to me and expressed friendliness as she waved me to come and sit at the table in the little garden. It was a lovely place with classical Baroque coming from inside the house, my Mom's favorite music. The woman came out with my coffee, plus a glass of freshly squeezed orange juice, and a homemade biscuit. I enjoyed the sun on my face, the music, the sanctuary, the treats and was feeling all the Camino was beginning to do for me. Sure my ankle hurt, my hip crunched, the blister burned, and my back was inflamed but at that moment in the little garden I transitioned into the next stage, the mind. I was excited, remembering what I learned at the pilgrim blessing from Terry, "The Camino will take you on a journey of three distinct stages. First, it will test your physical ability, then it will dive you into your mind, and finally open you to your inner spirit."

I hoisted my purple monster back on and was fishing into my pockets to pay when the woman came out of the house with a ladle in her hand. She reached for my shoulders, broth dripping from her tool, and kissed both my cheeks, wishing me a *Buen Camino* with a gentle push towards the gate. Slightly confused, I was walking away when a thought entered my mind, that wasn't a café, that was their home! I had just experienced unconditional love.

The seven hour day came to an end after 25 kilometers when I reached my goal of Belorado. It was strange coming out of the serene woods to face crossing a major highway to enter the town, but once through the gates, I was transported back in time. The streets of cobblestone were ever so uneven and narrow, a horse patiently waited tied to a post, and the old wooden buildings smelled their age. My senses were on high alert, I loved that place. I checked into a rustic inexpensive hostel that offered separate rooms. After the situation in Navette, I had a preference to sleep on my own. I unloaded my pack in my room, happy to lose the weight from my back, wondering if there was something I could be rid of to make it lighter. (No not the boots). I stripped off my sweat-drenched clothes in the shower and relished in the lukewarm water cascading over my beaten body. With a sense of pride, I admired the two bruises running down the front of my shoulders from the purple monster, a temporary medal of strength.

Tonight I would let my wounds air out, thankful for the sandals, and celebrate Canadian Thanksgiving in the town square. I sat down in the setting sun sipping on a cold glass of wine and watched the joyful energy of people around me. Int hat moment, I felt temporarily alone but found comfort in the observation and vibration of those around me. WiFi was available so I logged on Facebook and called my children to wish them Happy Thanksgiving and tell them how much I missed them, it was a delight to hear their voices. I messaged dear James, responding to his questions regarding the blister and wanting my current status. He was sincerely worried and was aware of how stubborn I could be. Smiling to myself, I assured him I would be fine, promising I would re-evaluate when to take days off. Tucked into bed that night, enjoying the energy of the purple painted wall behind my head (my favorite color), I drifted off to sleep with an empty mind.

BOB

I woke with contradicting thoughts the next morning. I felt refreshed from the night's sleep but my body was in great pain. Everything hurt and I had not even moved yet. I questioned whether leaving the wound exposed overnight was a good idea because it now burned intensely, like a million needles stabbing at it. Pulling the sheet back to have a look, it was red and angry ,with fluids seeping out onto the bed.

"SHIT, SHIT, SHIT!"

Then my attention was drawn to the very swollen throbbing ankle and hip. "Gotta get up Tess," I said, "Get the hell up now!"

I reached for Harold, my trusted stick , for the extra support to get my body moving. I re-wrapped my foot, got dressed, packed up, and limped downstairs to breakfast within the hostel. I sat down at the end of a long wooden table with three other women who all observed my uneven gait.

"You look terribly uncomfortable," one of them said to me with a kind smile.
"I am afraid I am, I have this stupid wound from a blister on my heel. Kinda thought I had it managed, yet now it is excruciating."
"Would you like me to have a look at it for you, I am a nurse," she offered.
"Wow, thanks but I have just wrapped it up for the day on my limited supplies. I took a picture of it before I dressed it if you want to see."
"Let us have a look then," she said, as she moved into a chair beside me. I pulled up the photo on my phone and handed it to her, she, in turn, shared it with the other two women, who were also nurses. The look on

their faces was not comforting.

"Have you been to a doctor?" the first woman asked.

"No, only a walk-in clinic a few days ago where it was lanced and bandaged."

"Well my dear you would be wise to have it looked at again, these sorts of wounds can run rampant. What happened to the stretched skin that lifted from the initial blister?" I dropped my head suddenly comprehending the mistake I had made, "I removed it."

"Right then the deed has been done," she said in a matter of fact way only a nurse can have when reprimanding a patient. "You must keep it clean and moist and do not let it dry out and, for the sake of yourself, go see a doctor as soon as you can. This kind of wound, especially out here, is a breeding ground for infection."

The three ladies stood to leave, wishing me the best, as I mumbled my thanks. Sitting alone my mind chastised how hard I had pushed my body, just like I always did, not making it number one. I heard the scrape of a chair on the stone floor and looked up into the warm face of gentleman with kind eyes. He sat beside me.

"Hi, I'm Victor, you don't mind if I join you?"

"Sure," I replied, "My name is Tess."

"I overheard some of your conversation with the other ladies, sounds like you're in a rough spot."

"Sort of. Though I got some good advice from them."

"Then I shall offer you none." He laughed, "My advice would be far more barbaric as I practice veterinarian medicine back home in Sweden."

"Just my luck, nurses and a vet coming to my rescue," I too laughed now at the impossibility.

We sat and chatted about our common desire to walk the Camino solo and the lessons we had been learning about ourselves. Together we smoked cigarettes, drank far too many coffees and shared relatable funny animal stories until the hostel owner hustled us out the door. It was wonderful to have a meaningful conversation with such an intelligent, thoughtful person, it really lifted my spirits. After Victor left, I quickly messaged James with an update, as a pending question was waiting for me. I told him briefly what the nurses had said and he responded with the insistence I take a day off. I persisted that I did not need to take a day off, I was fine, and that resulted in an argument and me signing off. He just didn't know me and my stubborn strength of mind.

I set out on that day consciously muting the pain as I felt the first real autumn chill in the air. The weather had been warm but finally fall had made an entrance. I had dressed for what I thought would be another warm day in shorts and a t-shirt. With the chill, I sought out every ray of sunshine on the trail for the fraction of the warmth it would offer. As much as I tried to ignore the pain coming from my foot, ankle, and hip, each step was tough and very slow. I shivered in my meager attire, not understanding why my mind would not command my body to stop and put more clothes on. I could only guess I thought it would get warmer. Others walked past me at an alarming rate, or so it seemed, and I felt frustrated at my weakness. The scenery was drab and did nothing to support my saddened mind and I was so very cold. After the longest four kilometers, I misread an arrow and found myself in the village of Tosantos, not on the Camino. I learned later that some *albergues* paint arrows to lead pilgrims to their establishments, in my case they had led me to feel utterly lost and very alone.

My mind was playing tricks on me and I felt I was falling into a delirious state. So much pain, so very cold, and all I could focus on was finding sunlight and warmth. I zoned out for a period of time and, not remembering how, found myself sitting on a bench in the sun along a paved road. So much confusion. I tried to operate my Camino app and it was impossible, my mind was shattered and the pain had finally become too much so I just sat there with my pack by my side. At some point, a huge travel bus pulled up right in front of me and my mind willed me to get on it but my body would not move so I watched it pull away. It was then I noticed a large man sitting just within my peripheral vision with his arms up behind his head basking in the sun. As I turned my sight towards him, trying to figure out how long he had been there, he spoke,

"Do you know where the Camino trail is? I seem to have taken a wrong turn," he said in a deep American drawl. I could only respond with a shake of my head. He then jumped up from his bench and came towards me slinging his pack on his back.

"Well my name is Bob and we are going to find it!"

He grabbed my pack and pulled me to my feet and strapped the purple monster on my back like I was a child, muttering about stupid arrows and not our fault. He took my elbow and began guiding me away from the bench, first to the left then the right until, finally, my body started to propel on its own again. My mind started to clear, it was like coming out of a trance, I could feel the ground again and I could feel the heat from the morning sun that was shining everywhere. I was once again connected to myself and my surroundings. Together, Bob and I found the proper arrows and we were once again back on the Way. I was sure then if Bob had not enforced his will, I would still be sitting on that bench. Once we established the correct direction, Bob did his best to stay with my slow pace but I could tell he was itching to walk on.

"Go ahead Bob, I will be okay now," I said, as I felt my inevitable strength returning.

"Ya sure Lil' lady?"

"Yes, all good now, you have done so much for me."

Touching the brim of his baseball cap to me we parted, him assuring me it was nothing and to take good care of myself. Once on my own again, regaining my bearings, I thought of the advice James had given me that I chose to ignore that morning. My stubborn side at most times served me well, but certainly not on that day.

Forcibly walking forward, looking at the ground, I took each uncomfortable step knowing there was an end somewhere. I climbed a small hill and looked up, for no reason, to see two familiar figures walking ahead. I cried out in relief as both Carlota and Jorge turned around and walked towards me.

"I screwed up, I am so hurting, I just don't know which end is up," I explained.

"We got you, Lil' sister, hold my hand. Here Carlota take her pack," Jorge said.

"No, I got it, it is helping me feel grounded," I explained.

Understanding my mindset, the two of them just walked slowly alongside me to offer support as needed. Just not being alone at that moment was all I wanted.

"Tess, enough is enough, when we get to the next village in two kilometers, we are going to make a plan for you," Carlota said. "Burgos is a large city, sixty kilometers away, it is time to have that foot looked at properly in a hospital."

Jorge piped in, "You are like a superwoman, but even she gets broken sometimes."

I knew then I had been defeated, not by the Camino, but by my own doing. A feeling of gloom washed over me, but at least I was feeling again.

THE BUS

After an hour, we arrived in the tiniest village. It had a small cantina where it was decided we would get me warmed up, I was still shivering from the morning chill. Jorge went straight to the bar to get me a hot coffee while Carlota got on her phone, rattling off in Spanish, organizing my future plans. It fascinated me that already in our few encounters, we had become so close and how the two of them embraced me with so much caring. I have never thought twice in my life about helping others and there I was experiencing it being returned. The gratitude I felt was so overwhelming that a single tear slipped out and rolled down my cheek. Jorge was at my side with my coffee and in a brotherly love wiped it away.

"Tess, you will be okay. Carlota and I will help you through this hurdle and you WILL finish the Camino."

He somehow knew what I was wrestling deep in my heart. The thing I had yet to even acknowledge—was I going to be able to complete this journey?

Jorge added, "You know Tess, Napoleon said, sometimes you have to lose a battle in order to win the war."

Hearing those words was when I realized that day was only one battle and I had to accept defeat, no matter how hard my ego tried to fight it, to be able to finish my Camino. Carlota joined us after completing another of several phone calls.

"Alright my friend," she said, "A bus about a kilometer away from here will be by in one hour. You will be on that bus and it will take you to Burgos. I have secured an inexpensive hotel there for the next four days, you will need the time to heal for sure. Once you have settled in, you can ask the front desk clerk, David, to call a taxi and it will take you to a state of the art hospital to have your blister looked at. This is what you will do. I have instructed David to make sure you go." She smiled at me with so much warmth and compassion. With my new understanding of a good defeat, I knew I would follow through.

The hour passed with us discussing failure and how, for mem catching a bus on the Camino felt like just that. I had convinced myself that, by doing so, I was no longer a true pilgrim. Jorge insisted I was not failing by addressing an injury, in fact, I was learning there was a limit to what my body could endure. That reassured me a bit and I knew it was the right decision in order to continue again.

When it was time to leave, Jorge and Carlota insisted on coming with me to the bus stop. The owner of the cantina, having overheard Carlota on the phone earlier, offered us a ride. I giggled to myself thinking they all wanted to make sure I got on the damn bus and didn't try to continue walking. Riding in a car, even the short distance, was already foreign to me having walked for so long. We passed Eva, the Dutch woman, resting on the side of the road and I slouched unconsciously in my seat, guilty.

We were dropped off at the bus stop on the side of the highway, if one could call it that because nothing indicated such, and there an older man asked in Spanish, "Were we too tired to walk?"

Carlota was quick to come to my defense, for she knew how fragile my thoughts were, and explained the situation whilst pointing to my leg. He nodded with understanding. Waiting for the bus, Jorge shared a funny story with me, a funny only on the Camino kind of story.

"We walked a short time with a man the other day that never seemed to be able to finish a sentence. Everything he said went something like this; 'Well, I saw this lady and…' then nothing else came out of his mouth." Jorge made funny faces as he continued. "But then 10 minutes later he would say, 'elegant.' The man did this on several topics and finally, I spun around and told him, 'come on man, you got to finish a sentence, you are sounding nuts.'" Hearing Jorge tell the story with his charismatic personality had Carlota and me in stitches. His expressions of exasperation as he relayed the frustration of the man was fantastic, Jorge was a great storyteller.

In the distance, we saw a big yellow travel bus approaching and Carlota started the frantic wave to flag it down. Her small stature and dark tresses bounced as she moved. I knew I had found my "pay it forward" person. I loved having that fearless little woman on my side, along with Jorge and his humor. After the bus stopped, Jorge picked up my pack as I hobbled behind him towards the driver. The man reached out to take my purple monster to put it under the bus, and I shouted out louder then I needed to,

"Wait, wait, can I have it with me please?"

The driver clearly understood me and said, "No," pointing under the bus.

"My journal, I must have it with me," I looked to Jorge pleading. The driver stopped, with whatever Jorge said to him, and I was allowed to reach in the pack and retrieve my beloved journal with all my memories of the Camino thus far. Then the driver pointed to Harold, for he too needed to go under the bus. Apparently my third leg was too ominous to ride in class, so on top of my pack he laid. I felt stripped of everything I relied on and it daunted me. My pack and Harold had become a part of my body and it felt strange without them.

As tt was time for me to board the engine-powered box, I turned to Carlota and Jorge,

"Thank you, my friends, I can't even express how much I appreciate what you have done for me today."

"It is nothing Tess, we love you, we are a family, born from the Camino," replied Carlota. "Hopefully we will see you in three days in Burgos."

"Ah come on you two," Jorge piped in, "It's time to go Lil' sister."

We embraced as though we might never see each other again, for we all knew it was up to the Camino, not us, if we would again. Out there, when you said goodbye to people, you said it knowing you may never see each other again and that was something I took for granted back home. I sat close to the front and could see out the big window my family waving like mad. As the bus pulled away there was Jorge jumping up and down while Carlota took pictures and I could only laugh. I cared so much for those two people whom I felt closer to than imaginable.

Silent tears rolled down my cheeks as I sat on the air-conditioned bus in comfort. It felt so wrong. My body had become conditioned to walking and I wanted to be walking more than I wanted to be sitting on that bus. I wrestled with the feeling of failure again but then remembered the words we shared in the cantina and fought the emotion. I gazed down at my Camino app on the phone and watched the blue direction line move at an impossible speed and thought of my fellow pilgrims out walking. What would take them two days on foot would be slightly over an hour for me. Carlota, having previously walked the Camino, told me those two days consisted of an eerie forest and many kilometers on the highway. I knew she told me to help ease my mind about taking the bus, and it did a little.

Coming into Burgos, the first thing I saw was the gothic steeple of the cathedral reaching into the sky over the large city. I had learned construction of the cathedral began sometime in the 13th century, at a time of French Gothic architecture, and was completed near the beginning of the 16th century. It was dedicated to the Virgin Mary. It's superb construction contained a famous collection of artworks, ornate choir stalls, tombs, and stunning stained-glass windows. I was instantly filled with excitement in hopes of exploring it once my foot was better.

Exiting the bus in the indoor bus station was an attack on my senses. I retrieved my pack and Harold, thanked the driver, and hobbled behind the others directly onto a busy city street. The noises were loud, the potent smell of a city, the chaos of movement, the harshness of concrete everywhere was more than I could handle. Gone entirely was the peaceful solitude of nature and I was forced to sit down and readjust my thinking to accommodate all the intrusion on my senses.

I looked at my GPS with the address of the hotel Carlota had arranged for me and could not make heads or tails of it. Smoke, I'll just have a smoke then it would all make sense. An Englishman stopped beside me to see if he could help. I showed him my phone screen and he looked just as perplexed as me and could not offer any information. He wished me luck and disappeared into the sea of people—there were so many people. Finished with smoke number two, I stood up and started doing the only thing I knew how to do, walk.

What a sad sight I must have looked struggling down the sidewalk with my uneven gait, surrounded by business suits and elegantly dressed women. I noticed up ahead a group of taxi drivers mulling beside their parked cars, waiting to be hired. I approached them and asked in my best Spanish for directions. They all eagerly looked at my GPS with the destination and began to explain and competitively point the same way. I was beginning to understand Spanish better and learned from the kind men that I needed to follow the curve in the street, turn right at a huge group of trees, pass three bridges, then turn left. I didn't think to hire one of them to drive me nor did they offer, it was the pilgrim way. Thanking the men, I started my long journey to find my room.

There was little relief in my leg from the bus journey, and the hard concrete footing was unforgiving, so my steps were labored and I had to rely heavily on Harold. The wind blew and, due to the higher elevation of the city, it was bitter and cold but still the beauty of Burgos unfolded before me as I travelled away from the nucleus. I was quick to fall in love with the evident history, unexpected river dotted with trees along the bank, and beautiful smiling people. Passing clothing stores, I caught the smell of the new cloth, wishing for a moment that I could buy a fresh outfit, but that would be silly. I caught my reflection in the window and was shocked how much I looked like a travelling gypsy. I was certainly not the fresh-faced poster girl for any outdoor store. I laughed at myself, I was doing it my way. Eventually I found my quaint hotel on a tidy street alongside the Arlanzon River. Even in a large metropolis city there is beauty if your eyes are opened. Making my way through the doors, I found David and, after stating my name, he welcomed me like an old friend. Carlota had arranged everything and he knew exactly what to do.

"No paperwork needed right now, let's get you up to your room and then I will call a cab to take you to the hospital. Your sister is very concerned for you."

"My sister...yes." I smiled at the reference and together we squeezed into the smallest elevator up to the third floor where David showed me to my room at the end of a long, narrow hall. He opened the door for me and stood aside handing me the key.

"Thank you, David."

"You are most welcome. Come down when you are ready and I will arrange for you to get to the hospital."

"Okay, will do," I said.

Once alone in the quiet room, I unhooked the purple monster from my back, dropping it to the floor, and laid on the bed. I was more exhausted than I had ever been in my whole busy life. Looking around the small room, I felt a sense of home—as it would be for the next four days—for the first time since I had left Him. The queen size bed was comfortable, the painted blue butterfly on the wall behind me brought serenity (Mom again, always with me), and there was even my own small bathroom with the tiniest bathtub, but it was a bathtub! I sent a quick message to James to tell him I was good and taking care of myself, minus telling him he was right, then closed my eyes and dozed off, the exhaustion getting the better of me.

THE HOSPITAL

When I became conscious again, my right hip throbbed as pain radiated all the way down my leg—it had been working overtime to compensate for my ankle and now it too was quitting. I limped to the bathroom to run warm water into the birdbath size tub thinking it could soothe my hip. I awkwardly maneuvered myself to be submerged, putting my legs up the tiled wall so I would fit in the tiny tub. It was then I could relish in the therapeutic water.

An hour later, I was in a cab with Edward, my English speaking driver, on the way to the hospital. During the ride, I learned he loved horses and his family and he learned how I got to this part of my Camino and why I was going to the hospital. He was kind enough to come in with me and sat me down in a waiting area while he spoke to administration to jumpstart my needs. Another angel from the Camino, who was gone before I could thank him.

I did not have to wait long in the bustling hospital before a nurse came out and announced my name. I used Harold to steady myself and followed her down a long hallway into a sterile examination room. Carefully she unwrapped the tired-looking bandage and, as the wound revealed itself, let out a small gasp. Gently she placed my leg back on the bed and muttered in Spanish as she left the room. I had a look and what I saw was not good. It was then I finally started taking it serious. It went deeper than I had realized, in fact, I saw bits of bone under the transparent layer of skin. The nurse returned with a colleague who was prompt in letting me know she spoke English.

"Hello, my name is Madeline. May I please have a look at your foot?" She smiled genuinely.

"Yes, of course," I replied. "I am so relieved you speak English, my Spanish is nowhere near good enough in a medical setting."

"Do not feel worried pilgrim, we will take care of you. Tell me how long have you walked in this condition?"

I was quick to calculate, not in days but kilometers. "About 80 kilometers."

"You are certainly courageous and a small bit, how we say, foolish." She smiled in understanding, "I am going to consult with the doctors and will return." With a gentle squeeze on my shoulder, Madeline left the room to return moments later with three doctors in tow. They nodded at me and turned their focus to examining the injury, sharing with each other their thoughts in Spanish. I was laying there helpless, not understanding what was being said, when Madeline appeared by my side again.

"I will translate for you what they are discussing," she began, "Your skin is damaged to a third level: it is an acute wound. The appearance of bone has them concerned, but the good news is you have no infection. They are bewildered how you have avoided that, and aside from walking, you have managed to care for it well."

One of the doctors spoke to Madeline in Spanish, which she was quick to translate for me.

"The plan is to perform a procedure of attaching layers of donated skin to protect the bone while your body heals itself. The doctors have concluded, having seen many pilgrim injuries, that it would be best for you to no longer walk the Camino but to go home and try again another time."

"Noooooo!" I cried out.

They didn't know what a fighter I was. Leaving the journey was unacceptable to me. I had come so far and to quit was not in my realm of thinking. The pang in my chest of not being able to finish was more than I could bear. After the procedure was done, I asked Madeline to translate my compromise with the doctor. He shook his head as she conveyed my thoughts with him. I persisted in my best Spanish—what I lacked in language I made up for in passion—I would do whatever he asked of me to make it happen. He finally agreed I could continue but only if there would be NO WALKING, only to the bathroom and back, for two days followed by one week off the Camino. I laid back relieved to have found a way to have hope.

Back in my room, I called down to David to ask if he could arrange for food supplies to be brought to me, he assured me he would. So for the next forty-eight hours, I was committed to resting in bed and healing. I was going to put every ounce of energy and faith into getting my foot back on the same plan as me.

UPRIGHT AGAIN

After forty-eight hours passed, I ventured back into the world, wearing sandals, for short strolls as per doctor's orders. My rehab consisted of walking from one cafe to another, stopping to drink cafe con leche and smoke Marlboros. I was surprised by how much the weather had changed with dark overcast skies and the looming threat of rain. The pain had become manageable thanks to the rest and medication prescribed by the doctor. The short walks felt good and I was feeling a new beginning coursing through me.

That night I enjoyed a simple meal and reminded myself how lucky I was to be sitting in the middle of this beautiful country. The commitment to healing was affirmed in my mind as the food nestled in my stomach and the wine softened my concerns. I had everything I needed. The Camino would embrace me back on the path, for I would walk with more appreciation of my life.

AN IDEA

On the third day of rehab in Burgos, I counted sixteen days since I had arrived in Spain, though it felt much longer. I felt a little frustrated that morning because I had been walking 24 to 30 kilometers a day but during my healing, all I could do was shuffle. Each footfall daunted me as I tried to find the step I had grown to know. What kind of pilgrim was I now? I missed the tribe of other walkers who I was sure were out in the crisp fall air negotiating rocky paths or climbing mountains. And me? I was sitting drinking coffee. Just a week ago I was sweating in a tank top walking my heart out. I wondered what my story would be in the end. I calculated I had twenty-one days left to cover 550 km, meaning I would have to cover 26 kilometers a day starting as soon as possible. My dreams attempted to shatter in a roller coaster of emotions as I wrestled with the reality of what I would be trying to accomplish. I went from bliss and gratitude to resentment of Him and my wound. My mind was in full force of playing a hand to challenge me.

The next day Carlota and Jorge would arrive. Via email, they had let me know they would be arriving in Burgos slightly behind schedule as Carlota also had problems with one of her feet. That afternoon, I walked a little further to find gifts for my friends to commemorate our small Camino family. I had clutched in my hand, on the way "home," three braided bracelets of different colors as the three of us were so very different yet entwined. When I was settled in my room for the night, I felt it was time to send an update back home. I had not posted anything for a few days and knew my friends would be wondering how things were going. At that point, I had not shared anything regarding my injury. I had insecurities as to what their reaction would be to my failure, which was likely only in my own thoughts. I posted the following:

"So here is the deal, my friends. I went to the hospital to have a blister looked at. In amazement, the doctors are shocked I have walked 80 kilometers with the ridiculous wound on my left heel. Well, it got the best of me and I thought it wise to have it looked at in the city of Burgos. Proud to say there is no infection but I do have a lovely Spanish skin graft. LOL. I have been told no Camino for one week! So I ponder this detour in my journey and have concluded this is my Camino no matter how it evolves. I am in this beautiful country with sites to behold. I do miss walking, and the so-called pilgrimage but define that word and I am still on it. It is my journey to learn lessons that are long overdue. There is no failure, just acceptance of what life brings especially when a lesson needs to be learned. I will listen to what I need to do, but no matter what I will walk into Santiago! I need to complete this for only myself. I am full of courage and an all-embracing heart that is mine alone. Please send healing thoughts so I can do this."

After I pressed send, I felt a new sense of confidence, proclaiming to myself that this had not been a failure, just a detour in my reconstruction. I read a comment that came through instantly from my friend Heather:

"To everyone in your life Tess you are our hero and truly made us all aware of the beauty of life around us through your eyes and strength, we have followed in amazement. Eyes wide open, with you while you 'Trust the Journey.' In my eyes, you succeeded when you landed in France."

Maybe that is why I walked the Camino, not only for me but to inspire others to live life. No matter how horrible the things you have to experience in your life, it will always be okay. For myself, having my heart shattered inspired the journey I was on, but it was becoming so much more. No longer was I walking a penance I thought I deserved, I was walking for my own freedom of spirit.

I was feeling stronger, so I ventured out to explore more of Burgos, finally, without Harold. My determination was powered up and my body followed. I found a yellow arrow embedded in the street, it made me happy and closer to the Camino. The farther I walked, the more I started seeing other pilgrims and I wished them all a *Buen Camino* as we passed, it brought the connection back I needed.

The air was chilly, and I was cold, so I bought myself a god awful Kermit the Frog green hoodie, but it was warm. Feeling good, I decided to keep walking and find the cathedral I was excited to see. Carlota and Jorge would be arriving in Burgos the next day and we had plans to explore the inside together, but at that moment I wanted to see the outside.

My steps took me to the cathedral square and the vision of the structure was more than I dared to believe. I felt the century-old history pouring out the endless towering turrets and sandstone walls. I stood in awe because it would forever be imprinted in my memory. Breaking the spell, I heard a distinct laugh coming from a tapas bar. I looked in that direction and saw Martha, from Rhode Island, and her troop. Promptly limping faster than I should, we locked eyes and reached to embrace each other. Wine was poured for me and we set about catching each other up on our trials and tribulations of the Camino. I gave a brief recount of my current situation and the waiting game I was forced to endure before carrying on.

Martha was quick to suggest, "Why not just rent a bike from here to Leon? My daughter and I are doing it. We have heard the way between Burgos and Leon is nothing but a long stretch of empty fields. Rather dull, don't you think?" She laughed from her belly.

"I never thought of that," I replied. "I think Martha, you have given me an idea. It would eliminate the pressure on my heel and I would be able to begin the Camino faster than I had thought."

With the seed of a possible solution manifesting, the two of us drank wine, told jokes and laughed at all our perils. I asked her if she was any closer to deciding about moving states to be with her boyfriend.

"Hell yes. I am going to do so when I return home. My lesson has been, life is too short and when a wonderful opportunity is given, just simply do it and be open to where it takes you." I smiled happily for my friend. "Your tattoo on your arm, 'Trust the Journey,' has been my constant reminder. I will never forget you or those words, so thanks girl."

I was floored my tattoo actually inspired another person and I promised Martha it would always ring true if she believed.

The evening came to an end and I was ready to get home and go to sleep. Bidding farewell, this time we exchanged emails because we both felt it might be the last time our paths would cross, sometimes you just knew.

As I trekked back to the hostel, I calculated I could get to Leon covering 180 km in three days on a bike. I would make it to Santiago in the time allowed and, more importantly, would still travel under my own power, which would be more rewarding than taking a bus or, worse yet, not finishing. Seeing a Camino shell on a building walking home, my thoughts drifted to the missing pack on my back and the direction of Santiago. The restlessness in my mind subsided at being shown a way, I would be going very soon. The Camino burned with fierceness in my soul.

DECISION MADE

"I am in Spain!" It was my first thought the next morning. The night before was the first time I had changed my bandage since being in the hospital and I was pleased to see the new skin fusing with my own. It was pink and healthy, aside from the top part, which was still some oozing, likely from too much walking the previous day. I decided to use that spot to gauge the healing process to completion. To see it dry and closed over would be the best outcome. But for the rest, "Way to go little body."

Dressed, I headed downstairs to my now familiar bistro under the hotel. Sitting inside by the window, I cradled my warm cafe con leche as the rain drizzled outside. The sky was dark and the wind was blowing but it did not affect my spirit, I would be seeing Carlota and Jorge that day. I spent time thinking about the bike idea to Leon and concluded it would be the best plan. I had spoken to James and he too thought it would be a brilliant solution to my situation. I was supposed to go back to the doctor that day but felt confident everything was healing just fine. I was too afraid of what he might say and gutsy enough to know I was far tougher than he would believe.

After I finished my coffee, I went to the pharmacy to mimic the supplies from the hospital and then went to the post office to mail home clothes I no longer needed. Back at the bistro, I ordered another cafe con leche but this time sat outside in the light drizzle to enjoy a smoke or two. I felt for my friends out on the Camino in the rain: Carlota, Jorge, Martha, Harold, Tatiana, Neil, and the professor, and hoped they were doing well.

It was definitely autumn now and I was happy I hadn't turfed my long pants to lighten my pack, I would have been screwed. A smiling pilgrim walked towards me and placed a card into my palm, wishing me a *Beun Camino*. I turned the card over and read the poem written on it;

> Why do I deal with the dry dust in my mouth?
> The mud on my aching feet:
> The lashing of the rain and the glaring sun on myself?
> Because of the beautiful churches?
> Because of the beautiful town?
> Because of the food? Because of the wine?
> No! Because I was summoned,

I was touched to be given this poem, written by an anonymous German poet. It summed up exactly how I felt and rang true to my presence on the Camino. What a wonderful gift. My mind and pen flew in the calculation:

Burgos to Santiago 503 km

St.-Jean-Pied to Santiago 815 km

What I had walked that far was 318 km

Burgos to Leon 180 km

Leon to Santiago 323 km

Allowing two more days of healing, if I rode the bike, with a rest day in Leon, I could be walking again by October 20th!"

That's it, I was going to rent a bike and cycle 180 km to Leon! Decision made officially, I paid for my coffee and slowly headed to the cathedral to reunite with Carlota and Jorge.

NO RULES

That day was a national holiday in Spain, known as Spanish Day-Columbus Day, and the city was alive with festivities. All of Spain was celebrating the day Christopher Columbus discovered America, when the world was proven to be round not flat. Columbus set sail from Spain in 1492 to find a western passage to India, instead he found America. He had been financed by the King and Queen of Spain, resulting in the national Spanish holiday.

I arrived in the cathedral square (where else does one meet up?) and saw Carlota and Jorge through the crowd at the same time they saw me. It was a loving, wonderful family reunion of smiles, hugs, and kisses. We had missed each other so much. Though it had only been a few days, the three of us felt like it an eternity. Arm in arm, we walked into a rather elegant restaurant to celebrate our reunion. Beautifully-dressed people sitting at white linen tables smiled at us in our pilgrim attire as we were led to our table. Once we were seated, our status as pilgrims allowed us to be treated like royalty. We feasted like kings on chicken, salad, omelets, wine, chocolates, and liquors, all compliments of the house! Jorge had given up on his "no wine" rule and we all toasted together, "No Rules!"

With the wine flowing, Jorge told funny stories in characters only he could do. Both Carlota and I were in stitches and tears from laughing so hard. Carlota told to me they had taken a cab the last 10 kilometers into Burgos as her blisters had been bothering her. We all smiled in the appreciation of "the other way" making an appearance again.

We exchanged thoughts about what the Camino was becoming for each of us and agreed it reunited us for a reason. That moment we knew, somehow, someway, we would walk into Santiago together. The conversation then went to my injury and the thoughts behind my bike decision. Jorge, as my newly adoptive big brother, wanted every detail of my barely laid plans until he was satisfied I had my head on straight. We agreed it was the best way to get me going again and would be easy to do as Jorge assured me the upcoming route was "mostly flat." They were going to stay in Burgos to wait out the weather, so Carlota insisted she would find me a bike. She was becoming the CEO of our needs.

Once lunch was finished, it was off to venture inside the Burgos Cathedral. We walked in hushed silence mesmerized in wonderment of the historic structure. Many saints lay buried deep within the catacombs. I sat in a pew that was five hundred years old and touched worn statues that had been caressed by so many over the centuries, it was humbling. My soul was overflowing with a connection to the history I had always craved to explore, and to share it with Jorge and Carlota, with mere smiles and nods, made it all very real. Pictures did not do the visual beauty surrounding us justice. We left the cathedral, after the self-directed tour of our experience, and agreed it was more than we could have imagined. Feeling peaceful, we strolled away together and stopped at a quaint, romantic looking hotel.

"Here is where we sleep," Jorge announced.
"Always a little romance on the Camino for us," replied Carlota gazing at Jorge.
"I feel such joy in watching you two. You are showing me a love of honesty and purity." I smiled at them. Hugging goodnight, trusting Carlota would advise me the following day of how and where to rent a bike, I continued home. I felt new love in the core of my heart. It made me smile and I felt my own broken heart coming back together, one piece at a time.

Coming around the corner to my hostel, my favorite bistro was still open with a solitary table and chair positioned outside under an umbrella. I wondered if it had been set out for me. Could they possibly know through my non-verbal visits that I would enjoy some wine in the rain? I took my seat and in moments the server came out with a glass of *vino blanco*, I wasn't surprised just delighted. There I was sitting outside in the rain, in Spain, I chuckled at this glorious solitary moment and felt so lucky.

THE MIRROR

When you are writhing in pain in Spain and no one can hear you, you are alone.

But I was no longer "alone," I thought at 3 am as I popped a couple of pain pills. Standing in my tiny bathroom, looking in the mirror at my reflection, I now knew I had me. I was home, home with myself. A rewarding conclusion to all the thoughts I had processed since the beginning of the Camino. It no longer mattered to me what structure I referred to as home, nor the things I kept within those walls, what mattered was my connection to my heart and soul, that's home. By not truly loving myself first, the love I had sent out, and wanted in return, only ever proved to be a temporary fix, until now when I was forced to love myself. By forgiving myself, I was now able to let go of all the things I had been blaming myself for,. The freedom that came from that forgiveness was allowing my heart to begin to beat in a different way, with its own independent rhythm. Justification of love from others was no longer relevant to life looking back at me in the mirror.

I knew the journey was taking me towards learning to appreciate the truth of who I was. There would be no more compromising for someone else. I had pushed so hard to prove to others for reasons I could no longer fathom. Maybe it was to Him to say, "Fuck you, look at what I can do without you!" as well as to the others that had hurt me in the past. I no longer needed anyone's approval, for all the confidence I needed was within me and I was breaking free. I knew exactly what I was capable of doing. I was strong, happy, powerful, free-spirited, and adventurous again. I had been summoned to the Camino for me and only me.

NOT TURNING LEFT

Walking out of my hostel the next day, feeling very much alive and happy, I turned right rather than left to my usual bistro. Instead, I went next door to La Mama. The instant I walked in, I spotted napoleon eclairs under the bakery display glass. By far the greatest chocolate treat in the world that I had come to love thanks to Carlota. It was interesting how a simple change of direction of where you are going can find you more of what you love, want, and need.

Delighted by the pastry, a warm coffee, and a smoke, I thought again of my early morning moment. I felt I had a better grasp of what my life would be like going forward. I seriously was trusting the journey, not just the one I was on but also the one I would experience for the rest of my life. I felt a new excitement for my future and my heart swelled with joy for all it could become.

The weather was changing and the sun fought to burn through the clouds as the air felt a little less chilled. I walked the few blocks to the dollar store to purchase some thermal tights, gloves, and earmuffs in preparation for my bike ride. With the purchases in hand, I took my steps thoughtfully back home, respecting my body's amazing healing process and stopping a few times to admire the beautiful city of Burgos that had been my temporary home. I popped upstairs to my room to unload my treasures and to check if there was a word from Carlota. Yes, she had sent me an email that she had made all the arrangements for the bike and would forward the information later. She signed off, "You rock Tess!" Back outside again, I celebrated the news from Carlota with a glass of wine at La Mamas in the company of the happiest little dog I had ever met. I relished in being 6000 kilometers away from home, in an outstanding foreign country with the freedom to drink wine and play with a dog. So simple and yet so very glorious. The man who owned the dog and I exchanged words in English and Spanish, having no clue word for word what we were saying but both agreeing the wine was good and the dog was happy.

I looked up then and saw my Camino family walking towards me, what a wonderful surprise, they had come to visit. Together we shared wine and "brother" Jorge and I bantered back and forth.

"Tess, you are a crazy Canadian you know?" Jorge ribbed me.

"I know, I know, but you...You are an equally crazy Spaniard man and don't deserve my Carlota. I am sure you will make her laugh to death if you are not careful," I replied.

"It is okay, it would be the best 'other way' to go," Carlota jumped in. Laughter spilled from all three of us as people who passed by smiled at our happy family.

"So Tess," Carlota said, "Your bike is booked and ready for you to pick up tomorrow at 6 pm. I spoke with David here at the hostel and have told him your plans, so he can house the bike overnight and assist you in any way."

"Wow Carlota, I thought I was a take charge woman!"

"We are so pleased to help you, Tess. You are such a light and inspiration to both of us."

"And we are family," I said as I reached into my pocket and pulled out the braided bracelets I had bought for the three of us. "No matter where we are on the Camino we will always be connected."

"Ah nice, my little sister, we will wear them with pride and a great love for you," said Jorge, as he promptly put his on after helping Carlota with hers.

"Together always, "we said in unison.

It was adios until we would meet again, maybe somewhere out on the Mestas, as Jorge and Carlota would begin walking the next day, while I waited for one more day to finish my rehab.

Later that evening, after resting my leg, I came downstairs to eat and realized I had changed alliances and once again strolled into La Mama. The lovely older woman behind the counter welcomed me with familiarity and, with Google translator, I was able to order myself the beef burger I had been craving for some time. The first night in Burgos, at the other bistro, Benza, I had ordered what I thought would be a beef burger but later discovered it was blood sausage. It was so gross in a dry-heaving kind of way.

While eating, I signed onto WiFi and updated Sarah, my daughter, on my plans, as well as James. It was nice being able to speak with a man easily. Back in my previous life, He never wanted to talk about anything that would result in digging deeper than the surface. I knew I would never engage in that type of relationship again, and was feeling the slow disconnection from Him. I was only sad for his journey that had led him down such a road, but it belonged to him alone. I did hope that one day he would awaken and see all the love and joy that he could have, but it wouldn't be from me.

Finishing my evening with a glass of wine, my thoughts drifted to my nights in the *albergues*. Since the onslaught of the blister, and the experience that one night, I had been self-caring in hostels. I needed the space and privacy to get through the physical and mental discomfort. I accepted I was not twenty anymore and believed the experience of the *albergues* was for the younger folk. The first few I stayed in were fun, though smelly, cold and crowded, I could live with that, all part of the experience. But then trying to sleep amongst the choir of snores only to finally drift off just as everyone else would wake up had grown old. I'd shared one *albergue* with five men who played a game of who could fart the longest and with the best smell! No matter how deep I burrowed into my bag, I could not get away from the stagnant aroma. Another time I was on a bottom bunk so low I would hit my head every time I tried to adjust my position. Oh, and I could not forget the time a couple had sex only a few feet away. Yes, it was the life of a pilgrim in *albergues*, I smiled to myself. In conclusion, I had the experience of that way but it was not for me. For a few euros more, the small private luxury was worth it

NELSON

It was my last day in Burgos. I would be back on my Camino the next day!

I lingered in bed for a while looking at the pictures I had taken thus far and noticed myself in the only same two outfits I had brought with me. I knew this to be true but seeing it in photos was eye-opening. At home, I sported many different fashion statements depending on my mood, yet I was the same on the inside. I realized looking at those photos, I was witnessing an evolution of myself through my own eyes. Daily I was looking the same on the outside, but inside I was rapidly changing, for the good. I could see a light in my eyes in the photos I had not seen in a very long time. It was a great way to start the day!

After getting dressed, I headed downstairs and returned once again to the first bistro for my morning cafe con leche. I felt an obligation to go there again on my last day, as their cigarette machine had kept me in Marlboros during my time off. The sun had returned, though not yet warm, it was physiologically bringing me good vibrations as did everyone else around me. For the fourth time that week, a local came up and asked for a smoke. I never minded but giggled because they always had a lighter so they were either closet smokers or budget conscious. I finished my coffee and walked across the street to the pharmacy to beef up my bandage supplies and anti-inflammatories. I had been diligent in taking care of my wound and was rewarded with less pain and visual improvement. The poor pharmacist pulled out every type of gauze, neither of us understanding each other, until I showed him my foot. Presto! For 2.90 euros, I had a massive supply of material and a strip of magic pills. As I waited for my purchase to be packed up, I read a poster on the wall about pilgrim injuries. At the bottom, it stated that only 20 percent of pilgrims who start from France make it to

Santiago in one walk. I smiled because I knew I would be one of the 20 percent with my fierce determination.

I had taken to counting my steps on my rehab in Burgos, more to stay connected to my walking. It was 189 steps to the pharmacy, 56 steps to Benza, 67 steps to La Mama. Tomorrow, I would pull out with everything I owned and continue. I would embrace my new thoughts, for it no longer mattered how I got it done, that was up to the Camino and I trusted in that.

Later in the day, sitting in the sun at La Mamas, nibbling on snacks that always came with afternoon wine, the feeling of excitement for the next days' departure engulfed me. I was setting out soon to pick up my new companion that would join my troop of the purple monster and Harold. The bike, which I had already fondly called, Nelson, after my friend James, would be waiting for me 2 kilometers away. Having a brief reality check of what lay ahead of me brought me a giggle. Firstly, I was going to ride a man-powered (in my case, a woman beast Canadian) bicycle for 180 kilometers, taking four days to do. Secondly, having already walked over 300 kilometers, I was thinking the bike would be pretty easy, BUT, I laughed out loud, I had not ridden a bike more than a block in thirty years! The feat would be done in sandals and with the harbored secret that my hip was still very sore. Thirdly, most days were becoming cold just standing still, so with the added wind as I zoomed along, I was seeing some serious humor in my future.

On my way to pick up Nelson, I got lost in the maze of streets trying to find the rental store. I finally stopped two approachable women to ask directions. Thankfully they knew some English and were happy to practice with me. The two ladies went out of their way to take me right to the doorstep. As we walked one of the ladies asked,

"Did you start in Roncesvalles?"
"No," I replied, "I started in St-Jean-Pied."
"You did the Pyrenees! Oh wow, that is fantastical."

They were both so impressed that it gave me a feeling of pride to acknowledge things that I had accomplished. There were no questions why I was riding a bike now, to them and to me, it just didn't matter.
"You are our hero of today." Hugging me at the door they continued on their way. What a delight it was to have had that moment with them.

Harold, Nelson and I met. Oscar, the agent who Carlota had conversed with, was wonderful to deal with and assured me Nelson would do the job.

Oscar answered all the questions I had and, most importantly, confirmed several times the ride would be, "Mostly flat." Pushing Nelson along the street back to my hostel, I thought I had better climb aboard and have a bit of a feel. I was a little nervous being surrounded by people and traffic but relied on the old expression, "It's like riding a bike," somethings you just never forget. Within a block, I had my mojo going and it was familiar and fun.

Back at the hostel, David helped me maneuver Nelson into the tiny elevator, back wheel on the floor and front wheel up the wall. To say it was tight would downplay the situation, but we laughed. With Nelson parked at the foot of my bed, it was time to repack for the Camino. I had to get all my belongings into the saddlebags provided, as well as the deflated purple monster. It took me several attempts but I finally got it and felt satisfied with my new skill. I contemplated dumping my boots because of the extra weight but I was determined to wear them again.

Lying in bed, after my last birdbath, I thought of my Camino family, Carlota and Jorge, and hoped they were doing well. I reread my pilgrim's prayer and reflected on the memorials I had seen of those who had died on the Camino. I had stopped at each one to pay respects and gratitude for those who shared the same passion I had. I closed my eyes and promised to all those souls I would carry them with me the rest of the Way.

MOSTLY FLAT

Waking up briefly in the middle of the night, a full moon shone through my tiny window, casting a guiding glow over me. I smiled, dozing off again until the bright, beautiful sun replaced the moon to herald the start of a new day. I asked the universe to please give me an easy ride with no pain.

I layered up with all my warm clothing for the 7 am start and opted to bounce Nelson down the three flights of stairs rather than struggle with the elevator. I checked out and said goodbye to David with many thanks. He, in turn, wished me a *Buen Camino* with assurance I would make it to Santiago. Once outside, embracing the cool air, I pulled down my ear muffs, pulled up my Kermit the frog collar, and began peddling away from the neighborhood that had been my sanctuary. I had completed the physical stage of the Camino in that beautiful place.

With Carlota's verbal directions in mind, I followed the river to a spectacular stone bridge and it was there I found the yellow arrows. Once back on Camino I was deliriously happy! The sun was shining on my face and the awareness I had conquered a huge setback was all the power I needed. Nelson rolled along smoothly as Harold was strapped across my saddlebags feeling, I was sure, rather obsolete. My right hip was not very comfortable so I put my focus on making my left side work harder. With less body weight on the injury, I had to retrain my brain to use that side again. I had small waves of guilt as I passed pilgrims walking, noticing looks similar to my own before when bikers had passed me. If their thoughts were anything like mine, they'd be thinking, "You got it easy," but I knew now that I was simply doing the Camino anyway I could. Smiling as I rode by, I wished every one of them a *Buen Camino*, some responded some did not.

I needed to be mindful of the arrows now, as they sneaked up faster on a bike and could easily be missed. What fun I was having! Memories surfaced of being a child whizzing around the neighborhood trails, a favorite pastime for me. My body warmed up, my eyes were wide open, and I loved feeling the wind on my face.

After I had completed 21 kilometers, I stopped to take a break. What would have taken me all day to walk I accomplished in half the time on the bike. Sipping a cafe con leche and eating a biscuit, with Nelson positioned against a wall, I laughed at both Jorge and Oscar for insisting the ride would be "mostly flat." The last 10 kilometers had certainly proved them both wrong as I had encountered a hill that even the best cyclist could not have peddled up. Not only did I have my gear to get up but also the added weight and awkwardness of Nelson attached to all that gear. I had shook my head in disbelief at how difficult it was to push a loaded bike up a 10 percent grade. After what had seemed an eternity, I reached the top and was greeted by an expansive meadow plateau with one solitary tree in the far distance. Happy to be on level ground, I climbed aboard Nelson and peddled across the terrain, finally mastering the gears to accommodate the small ups and downs of the trail. Reaching the abrupt edge of the plateau, I braked hard. With all the steps I had taken on the Camino, one thing I had learned was what went up must always come down. Before me was a descent I was sure Mother Nature had devised for the BMX Olympics!

There was no bloody way I was going to ride Nelson down that hill, I would have certainly ended up "tits over backward." Off the bike again, (mostly flat...my ass) I guided Nelson down using the hand brake so as not to have him run away from me with all the weight he was carrying. There I was pushing a bike while carrying my walking stick and boots. Instead of my Camino aids aiding me, I was aiding them. The irony was not lost on me.

Break over, I was back on Nelson, covering the level pathway at a nice clip when up in the distance I saw a familiar silhouette, Carlota and Jorge hand in hand. I pedaled as fast as I could shouting, "Mi familia. Hola, mi Familia!" The happiness of being reunited once again was wonderful. There we were, not far from the middle of nowhere, finding each other. In life, we take for granted people we love and care about, assuming we will see them all the time. The Camino teaches you there is never a promise of tomorrow, a lesson the three of us had learned very fast. Together we walked, Jorge pushing Nelson for me, to the next village to have a meal.

"So Jorge, the bike ride so far has NOT been mostly flat. I am not sure what you Spanish folk think is flat but I can sure tell you what isn't." I snorted as I told them of my experience.

They both tried to stifle their snickers. When it could no longer be contained, we all broke into fits of laughter.

After we finished our meal, Carlota got on her miracle-making phone to arrange accommodations for that night for the two of them, and, thoughtfully, for me as well. She insisted on making sure I was comfortable, warm, and focused on taking care of myself. That woman was becoming a sister to me in every sense.

Carlota was busy on her phone and Jorge had gone inside to pay the bill when I noticed an amazing woman sitting on the curb not far from me. She had a beautifully kind face and I could tell, even sitting down, she was tall, but what really stood out was her hair. It was shaved from her neck up to her ear and tied in a ponytail of dreadlocks cascading down in every color. Dressed in a similar fashion, she had piercings on her ears, nose, eyebrow, and lips. The vision all together was magical and pleasing.

"Heeeeeey, my name is Rainbow just in case you hadn't guessed yet," she said to me with her illuminating smile.

"Wow, of course, it is!" I replied. "Sorry for staring but you are stunning and your appearance makes me smile."

"Well," she said in a rich Swedish accent, "That is my intent. I am walking this Camino thing making people look at me and hopefully feel what you did, happy."

"You certainly succeeded with me," I said. Then, pausing for a moment, I asked, "Are you happy?"

"Ah clever question, I knew there was something different about you too. I'll tell you, I have been sad for so long in my head, so I am walking trying to find my happiness. As people pass me or, like you, really see me, for the most part, I get smiles and good feelings. I am taking energy from the moment and adding it to mine. I am using it to help build a reservoir of joy, fueling my soul."

"Then I will easily give you some of mine because you have given me great pleasure in meeting you."

"Yes, thank you, the feeling is mutual," she said, "Now I am off to find more."

Unfolding her long legs, she strode off with a wave. I watched that rainbow beauty disappear around the corner leaving a stream of colors in her wake.

Turning back to Carlota and Jorge, who were both smiling having witnessed the tail end of the encounter with Rainbow, Jorge said, "Little sister, you attract the most interesting people."

We walked away from the small village together for a while before Jorge sensed I was ready to roll again and encouraged me to get going.

"I am in need of alone time with my girl," he said, winking at me.

"Of course you do big brother," I replied with a smile.

"Let's see if the Camino brings us together again in Castrojeriz tonight," Carlota piped in.

With hugs and kisses goodbye, I was off, leaving the two lovebirds on their own. Their times alone walking was creating a beautiful bond, as they were closer in love every time I saw them.

The next part of the trail held its own challenges as it was very narrow and rocky with huge muddy hollows. It would have been easier walking. Instead I had to precariously maneuver Nelson through and around the obstacles, careful to pass other pilgrims respectfully when I came upon them. Gratefully, back on a small paved road, I came upon the ruins of San Anton Church. I had to stop to have a better look. Majestic were the remaining ruins of blackened stones after centuries of weathering. I found a small carving under the archway, barely visible, the date, 1146. I could hardly conceive how something I was standing in front of could be so very old. There was a small opening in the wall and I could see prayers had been left by others for people they loved. I tore out a paper from my journal and wrote the names of my three children, as well as Jorge and Carlota's followed by a wish for them to all be happy. I placed the paper in the opening, hoping the two of them might find it as they passed through, like a message in the bottle.

Back on Nelson, weaving my way through sheep crossing the road, my neck and arms started to ache. I was grateful I could see ahead the town of Castrojeriz popping up out of the rolling landscape. I had just completed 60 kilometers on a bike in one day and was happy to soon be stopping.

As I settled into my hostel, arranged by Carlota, I decided this was my favorite sleep place so far. It was cozy and inviting in my small room and the window, with geraniums in a flower box, allowed a perfect view of a deserted castle on a hill. That hostel was located directly on the Way and the entrance to the town, so I went outside into the garden with my *vino blanco* to welcome my family.

When they did appear, I promptly invited them over for a much deserved glass of cold wine. I asked them if they got the wish I left them but, unfortunately, no they had not, as they had been too overwhelmed with the ruins to notice. They were both very touched by what I had done for them. We chatted then about the impending moon scheduled to rise the following night. Knowing we would not be together, I wanted to share my passion for the Hunter Moon that would grace the Spanish sky.

"The moon on this night is when the veil between worlds is the thinnest. It is a night to open your senses and experience spiritual growth. If you are open, you will feel a shift in the energy around you if you are paying heed. The air will be crisper and nights become longer as we begin transition with the turn of the Wheel of Life," I explained.

Carlota's eyes widened as she asked, "I feel I have been seeking for some time fresh new energy, a fresh feeling of my Camino. I have been carrying parts on my walk of my previous Camino that I want to let go of. Would this be the time? How do I do this?"

"Yes this is a great time, interesting how I knew to share with you both about this moon. I know this because of my personal spiritual practice of Wiccan."

"What is Wiccan Tess? What does it mean?" skeptical Jorge asked.

"Wiccan belief is a strong connection to Mother Nature that is so powerful and the world we live in. I respect the elements of earth, air, fire, water and energy, along with the balance of directions, East, West, South and North. I celebrate and honor in private rituals the changing of the seasons, the solstices of winter and summer and, my personal favorite, the stages of the moon. It is a practice that only asks you to continually grow and change, just as Mother Nature herself does."

"Yes I see all that in you Tess, you are exactly that," Carlota said as she reached for Jorge's hand. "We are both so thankful that the three of us have connected. So I can draw on this moon in my own way to fill me with the new energy I so desire?"

"All you need to do is believe it is there, then feel it, be open to it. It will create a growth in you for that fresh feeling, just like the growing love for Jorge you now feel."

"Tess, your words of wisdom have sent shivers of warmth through me," Carlota said.

"I am always happy to share what I know to those who are open and want help to fan the flame that burns in our souls. Some may choose to ignore that flame, or better word passion, but not I, or you."

"Okay girls, now we sleep," chirped in the not so serious, but affected Jorge.

We said our goodbyes with plans to meet for breakfast in the morning. As I watched them leave for their hotel, I felt a whoosh of love for those two beautiful people. I had given them a gift in return for all that they had done for me.

HUNTER MOON

I woke up the following morning from a restful sleep. The physical stage of the Camino was now very much a part of me. I knew I would be feeling some sort of discomfort all the time and that would be my new normal. The wound was healing, with careful management on my part, and the pain in my right hip was submissive with the anti-inflammatories and rest breaks throughout the days. Riding the bike had awakened muscles I hadn't used in a long time and I could feel new strength building.

I packed up Nelson and set out to meet Jorge and Carlota for breakfast. It was a feast of fruit, yogurt and homemade slices of warm bread, which we indulged hungrily in. Carlota and I went over my route to Leon, figuring out where would be the best places for me to stop for the nights. I was planning to take a day off in Leon, to explore the city on foot in preparation for walking again, and Carlota pointed out that we might be able to reconnect there. Though wary that the Camino had its own plans for us, it would be nice to walk into Santiago together. To share the final destination together would certainly put our relationship at a level of intimacy, as it would be a very personal experience.

Too many goodbyes, I thought, as I peddled Nelson out of town after breakfast. I was now concerned I wouldn't see the two of them again like the others I had met. All I could do was put trust in the Camino and the journey, as it did seem to be reuniting us for whatever reasons we hadn't learned yet.

After ten minutes along a pleasant flat path, I braked hard and looked up in astonishment. "You have got to be kidding me," I said out loud. In front of me was what appeared to be the mini Pyrenees with a sign assuring

1050 meters elevation with a drawing showing how steep the grade was, 12 percent. Taking into account my full saddlebags, stupid boots, and Harold all weighing in around sixteen kilograms, plus pushing a bike awkwardly, I began the steep climb. I was losing faith in any concept of "mostly flat," and it was clear that riding a bike was no easier than walking. Steady, slow, stop, stretch, breathe, switch sides, repeat and repeat and repeat. Just when I was reaching my limit, somewhere near the top, a young man came up from behind and started pushing Nelson, which almost put me in a jog. I couldn't help but laugh almost tripping over myself to keep up. Reaching the top, I thanked the pilgrim for his help, happily rewarding him with the cigarette he asked for, then climbed aboard Nelson and was off negotiating the uneven mountain path. Any wrong maneuver could take me out and some of the pilgrims were not willing to move to let me pass. I pushed up more hills followed by insane downs that removed all the judgment I'd felt previously towards cyclists. I developed bruises in unmentionable places, my arse was rubbed raw, and my neck and shoulders ached from the unaccustomed angled position. I was doing my best to embrace the experience but I was really looking forward to walking again. There was no cheating on the Camino, it would not allow it. With 30 kilometers left to go, I was still hoping that maybe then it would be "mostly flat."

It turned out it was for the most part but the going was very slow as the day wore on. The path, sometimes narrow, was uneven and littered with large rocks that I had to maneuver around while the wind blew fiercely against me. I made a bad decision to go off the bumpy trail to ride in the smoothness of a grain field, only to get stuck in unseen mud, which resulted in pushing Nelson back two kilometers to get back on the Way.

Exhaustion was overcoming me and all I could do was focus on the front tire going round and round. Every injury, both recent and historical, screamed profanities at me. I was reminded of the day I hobbled into Najera. With 59 kilometers behind me, the horizon finally gave way to the encouragement I needed to finish my day, the town Carrion de las Condes, where I would spend the night. With relief, I quickly found my hostel, parked Nelson in storage, and staggered upstairs to my room. I was too tired to even go for my customary walk about town. Instead, I stripped off my filthy clothes and submerged myself and my laundry into the floor shared tub. It was there I honored the Hunter Moon and opened my heart even more to what life was going to offer.

THE MESETA

The next morning, as I moved each limb checking for pain, I smiled and expressed gratitude for the healing Hunter moon and hot bath. I hoped Carlota too had received what she needed from the energized moon.

I had coffee in the town's square then ventured into a store and found a solitary silver necklace with a crafted Camino shell hanging from it. I purchased it to commemorate the power of succeeding at the previous day's ride. I looked down at the bracelet from Tatiana on my wrist, which represented the completion of an arduous walk days ago.

Packed up and ready to go, I looked with disgust at my boots hanging on either side of my saddlebags. They had caused me so much grief, yet here I was giving them the easiest Camino any pair of boots had ever done. Then I had an idea. Inside those boots were the insoles I had bought in Pamplona that were cushy and soft and the seat of my bike was not. I reached inside the boots and pulled the inserts out and promptly stuffed them down the inside of my tights to sit on. I could just imagine what I looked like, I smirked, with a tail trying to break free from my tights, but I no longer cared, it was all about comfort.

An hour later, I came upon the beginning of the Meseta. Stretched out beyond what my eye could see was a very large expansive flat plain void of anything vertical. The acute second stage of the Camino, the mind, was before me and I could visually understand why. I had heard the distance was long and the landscape empty with nothing to distract my thoughts. The contrast of the dry, yellow, endless plain touching the brightness of the great blue sky at the far horizon was stunning. Having started the transition of the second stage when I left Burgos, the Meseta would challenge my new found thoughts. In the hecticness of our daily lives, we push the strains and uncertainties that we can't find a way to resolve to the back burner. We unconsciously harbor our feelings of fear, doubt, sorrow, and regret because they are either unpalatable or intractable. But after hundreds of kilometers, my body had been broken down and reformed, and in that breaking, I had entered a place where my emotions were free to seep out. I embraced the Meseta to create an opportunity to breathe deeply, gain perspective and heal from the negative thoughts I had carried for so long. It was there I learned to think a thought from beginning to end and how I finally understood how to let go of notions that no longer served me in a positive way. The sight of other pilgrims I passed, either brimming with joy or crying tears of release, created an environment of acceptance of both ourselves and the things we could not control.

Halfway across, a small village emerged in a dip, like an oasis on the flat terrain. It was a good place to stop for a cafe con leche and a smoke. I was eager to write in my journal as my thoughts were flying and writing would help me release them. Finding a small café, I sat down at an outside table and stifled a laugh. I couldn't wait to tell Jorge that at last, I found "mostly flat." A young Korean boy came walking towards me from the opposite direction from where I had arrived. Perplexed as to how that was possible, I smiled at him and asked,

"Are you lost?"

Speaking in precise English he replied with a huge grin, "I left my *albergue* this morning in the dark and certainly did get lost, turning left where I should have turned right."

I grinned, remembering my same mistake before Los Arcos.

"I rerouted myself 13 kilometers ago and I am here now."

"Are you on the Camino alone?" I asked.

"No, I have friends but I have not seen them in days. I am not sure if they are ahead or behind me. We all just finished studying in France two weeks ago, receiving our degrees in physics, and yet we can't seem to figure out this Camino!" We both laughed at the irony of the situation.

"Well, you certainly are a smart man accomplishing a degree in a foreign country," I said.

"Not really," he replied, "I got lost."

I enjoyed the day's ride and only had 10 kilometers left to go to Sahagun, home for the night. I was peddling along in the relative comfort of both body and mind. I had clocked 120 km on Nelson at that point and was feeling pretty formidable. After the end of the Meseta, the Way followed the same straightness, but along a busy highway. I was quick to miss the beauty and tranquility from the vast plains I had just ridden through and could see the Camino was changing before me. The forests, sloping hills and fields would come again though it would be different as I traveled through the different regions of Spain. Soon, after a day or two off in Leon, I would be walking again. I was feeling so very ready. Silly thoughts went through my head about buying Nelson and shipping him home, for he—like Harold my walking stick and the purple monster—had become a strange attachment. But a returned email from Oscar at the rental company assured me this was not possible. I laughed thinking how far the mind can roam on the Camino.

ROUTINE OF SORTS

I arrived in Sahagun earlier in the day than expected and quickly found my hostel where I tucked Nelson away and set out to explore the town. I had energy and legs that felt the need to walk again. I was quick to discover it was a strangely laid out town, with streets that zigged and zagged, continuously bringing me back to the same spot. That same spot had a nice bistro so I decided to enjoy a coffee and do some people watching. I heard a familiar voice a few tables over and looked up to see Patrick, the tattooed Irish man I last saw from the party night in Estella. We greeted each other with a warm smile and a brief catch up, which entailed his own slow down on the Camino, pointing to his bruised and swollen bare feet. He was having a tough go with a couple of broken toes and a heat rash that had opened his skin so he was now walking the Camino bare footed. Pointing to a pretty girl sitting at the table he had come from, Patrick proudly announced he was longer alone. Camino love can be so much of what one needs for encouragement to keep going. It was wonderful to see him again, we hugged goodbye, and I headed back to my room in the early evening chilly air.

Fall was encroaching the Camino and I needed a hot shower and more clothes to feel warm. I checked my email and a message from Carlota assured me they were chugging along. She told me they had taken a cab to the village of Formista because her feet were giving her trouble again. The two of them had celebrated the full moon and both felt a change and a great deal of positivity. There was also an email from Neil and Tatiana, who had also taken in the Hunter moon as they had thought of me. I was very proud of my friends.

Bundled up the best I could with my meager wardrobe, I set out in search of a pharmacy to restock my bandaging supplies. The wound was healing with my diligent care, the donated Spanish skin, and Nelson. I was full of confidence that I would be able to walk out of Leon. With my bag of goodies from the pharmacy, I stumbled upon an Asian dollar store packed with every imaginable item. I could barely squeeze through the aisles. I purchased new maroon-colored tights, as my current pair had worn holes in the unmentionable area, a new notebook, and a polar fleece scarf to ward off the cold air that was coming. The expense was minimal but the comfort immeasurable. The next stop was a tobacco store, I was now a full member of the Marlboro club. I had resigned to not quit smoking out there, as it was the comfort I needed. Relishing an evening glass of wine in the square, I realized that by not staying in *albergues* my social life was nonexistent. However, I was truly enjoying my time alone, falling in love more every day with the woman I hardly knew before the start of the Camino.

Back in the room for the night, relaxing on the lumpy, yet cozy, bed, I felt the wave of appreciation of being in a foreign country and learning to live within it. Everything that I had experienced to date was nothing I could have imagined when the plan was first born. The array of people from all over the world, the accents, the food, my Camino family, being a pilgrim, the aches, the injury and the pain, both physical and mental, were all a part of the adventure. The feeling of Santiago getting closer, as I was just over the halfway mark, breathed into my awakened mind.

There would be so much more to experience, more kilometers to walk. I had come to accept this as a new way of living. The routine I had established was to wake up around 6:30 am, repack my bag, drop key off, search for coffee, smoke, walk six to eight hours, find a place to sleep, get Camino passport stamped, new bed, explore a new town, two glasses of wine, eat, smoke, journal, shower, hand wash my clothes, hang to dry, put on the second outfit to sleep in and wear the next day, input WiFi password, check messages, send replies, dress wound, sleep, and repeat. The changes I looked forward to every day were the scenery, the people, filling pages in my journal, the unexpected terrain, the history, the weather, the new towns, and my thoughts. It was a wonderful life.

There were times I would miss "home," but other than my kids, I no longer knew where or what home was. Back in my memory, where I had settled those thoughts someplace that didn't hurt as much anymore, I thought of the house I lived in with Him. The Camino was giving me the assurance I needed of independence and freedom from attachment to any one person, which I had become aware was a missing link in my adult life.

INVENTORY

The next day, I was happy to see the promise of a blue sky. It would bring warmth and the feeling I get when the sun shines. Sipping on my morning coffee with Nelson all packed up by my side, the sun was sneaking its way through the buildings, allowing for splashes of light throughout the square. The earthly colors bounced off the ancient architecture, enticing me to paint a vivid picture in my memory that no photo could capture. My coffee here was served in a glass cup rather than the traditional ceramic I was accustomed to. The new region of Spain was bringing subtle changes and the glass cup provided an even better tasting coffee.

I was journaling daily, both morning and night of my travels, and it was giving me such pleasure. That day, I had to retire my first journal, which was full of memories and falling apart, and start the new one I had bought the night before. I remembered the day Sarah and I bought the first journal at Walmart back in Canada, debating for some time which one was light enough to carry and here I was starting the second one. Often I would look behind me and see that journal peeking out from my pack, it was the most treasured item on my journey along with my family bracelet and silver Camino necklace. With so few things in my possession, each item had a purpose and I would often unpack everything to reassure myself it was all there, even though I knew it was.

How easily we take for granted our belongings scattered throughout our homes. How unconsciously we continue to purchase them, whether we need them or not, adding to our stockpile only to forget we even have them. Then one day through a catastrophic loss, whether by fire, theft, Mother Nature, or a break up, you are forced to realize those things are not who you are. In my case, five months prior, I was forced to condense many years of who I thought I was into the bed of my pickup truck in one day. It was a devastating time for me. I had to repeat the process again three months later to accumulate the cash for the Camino. That time was a bit easier as my heart knew I had to let go of the past with Him so I could truly start moving forward. All I had now, when I returned to Canada, a 1993 RV trailer, dubbed Dragonfly, that somehow I was going to make a new home for myself in.

On the Camino I owned:
1 purple backpack
1 purple sleeping bag
2 journals
Mementos from people I had met
iPhone and charger
Gloves and a toque
3 panties
2 pairs of socks
2 t-shirts
1 tank top
1 pair of tights
1 pair of yoga pants
1 bright green fleecy
1 raincoat
2 pens
2 dollar store reading glasses
A scarf and a buff
Sandals
Hiking boots
First aid kit
Toothbrush and paste
A small bottle of body wash
Hairbrush and elastics.

Not ever in my life could I list all my possessions from memory like that. I had an appreciation for what I did have, carrying only what I needed and loved. Items I desired to purchase or things given to me went through my mind first in a thoughtful process before I would add them to my pack. It was a lesson I learned that I would practice for the rest of my life.

The two most asked questions on the Camino were, "Where are you from?" and "Where did you start?" Most people assumed I was from the United States and I would simply explain I was from Canada, and all nationalities always would apologize for the assumption. I was never sure why they did, possibly the global stigma that comes from being an American. However, I had met some fantastic and kind Americans such as Neil, Tatiana, Martha and, of course Bob, as well as some grumpy stone-faced Canadians, so as far as I was concerned that blew the theory all to hell. The second question I would answer, "St. Jean Pied" and I would be congratulated and admired. I always thought inwardly, where the hell else would I have started The Way of St. James? Google told me it started there. I was discovering the further into the Camino I went, the less pilgrims had started in France, opting to begin in Pamplona, and fewer in Roncesvalles after the Pyrenees Mountains.

I gazed at the tower clock in the square, it was past 10 am. The early morning rush hour of pilgrims was long gone. I leisurely finished my coffee, surrounded by the locals starting their days as relaxed as I was. Jorge would have been proud of me experiencing the Camino the "other" way.

Once I was off, the first 10 kilometers flew by on a steady path with the sun shining in the crisp air, it was perfect. Crossing a pedestrian bridge over a serious interstate highway, I was being followed by the first dragonfly I had seen in Spain. I had a passion for these amazing creatures and for what they symbolized. The dragonfly, in almost every part of the world, symbolized change and change in the perspective of self-realization, with an understanding of a deeper meaning for life. I was amazed at how it found me at that moment and grateful for the little gifts that I was receiving.

Biking through the tiny village of Bercianos, I was drawn to the sound of classic blues coming from a small outdoor bar. I never liked to miss an opportunity to enjoy live music, so I parked Nelson and sat down at one of the tables outside. The energy of the place was as wonderful as the music that swirled around me. Sipping on my cafe con leche, I looked up and who should I see striding towards me smiling ear to ear? The young Korean physics major from France. To see a face you recognize, along with the immediate warmth exchanged, is the best form of unmasked human connection. It was nice to see he had a friend in tow.

"I have found one of my friends but only by getting lost again," he said to me.
"That is fantastic, but really, you gotta stop getting lost," I laughed.
"I will certainly try from now on. I have told my friend about you and how I would never forget your smile."
"And I would never forget how you can laugh at yourself," I replied.
"What choice do I have? I make mistakes that I just laugh at, much better than anger."

"Yes, I have learned a valuable lesson from you," I said.

"I also have from you Tess, your fierce independence reminds me to embrace life. Even when I am lost," He laughed. With that he hugged me, whispering a good Camino in my ear and was off. I shook my head with a smile at how I was in constant amazement of the journey every day.

With my fill of music, coffee and smokes, I was back on the Camino. As I biked along, I kept seeing signage, "Del real Camino." I had to laugh, wondering if after 450 kilometers I had been doing the fake one. The graffiti I passed on tunnel walls was full of positive quotes: "You don't choose your life, you live it" and "The highway from hell is over, the highway of flatness, the Meseta is behind you, life awaits." Words of courage coming from fellow pilgrims were exactly what fueled inspiration and connected us all.

After happily biking through the next village in the early afternoon sun, of all the eyeballs in Spain, a very mean stinging bug flew directly into mine. I slammed on the brakes as I was instantly blinded, almost crashing Nelson. I desperately tried to wipe my eye of the offending bug with my finger but it only proved futile. I sat down at a bench and took a selfie to see where the horrid creature was. Sure enough, he was lodged in the corner of my eye setting up house. I reached into my first aid supply and pulled out a gauze square, watered it down and as tactfully as I could, blotted and twisted until the offending bug appeared on the white square. Unfortunately, the victory was short-lived as some damage had been done, my eye would water for some time, but at least I could see better. My nose, apparently connected to the eye, had started to run like mad, and I had to blow it on all I had, my shirt sleeve, my glove and, finally, just into the air. Onward.

I came upon a small, old stone bridge on the trail crossing over what could only be described as a babbling brook. I stopped and looked around and discovered I was in the middle of an enchanted forest. It was like being in a fairy tale from the books I had read as a child. I felt the magic in the air. The trees were various shades of greens and browns in celebration of fall, the birds were singing songs of joy, the soft wind blew through the leaves, and the water rolled over the stones in the brook. Altogether it created a crescendo of nature at its finest.

I arrived in Manzilla six hours after the start of my day and decided to change my plans of spending the night there because I was feeling strong and it was "only" 20 more kilometers to Leon. But first, I stopped for wine, remembering Jorge declaring there are no rules, and indulged in a glass of feel good. My thoughts went to the encounter I had an hour before finding

Eva, my Dutch friend. I was passing a pilgrim rest area at the side of a road and to my delight saw her sitting there. I promptly stopped Nelson, dumping him in the dirt, and jogged over to her. Eva had become the longest recurring person I had met on the Camino and I would not miss an opportunity for yet another encounter.

"Hello, Eva!," I called out as I approached her.

"Oh my God! Tess, it IS you. I saw you biking towards here and I thought that looks like Tess, but no, it can't be, she's not on a bike."

"I know right? But yes, here I am and it is so good to see you," I said.

"What has happened to put you on a bike? Your ankle?" she asked, remembering me hobbling out of Zubiri.

"No, a ridiculous blister laid me up in Burgos for a few days, this was my only way to keep going and heal at the same time. I plan to start walking again when I leave Leon. Never mind me, how are you, my friend?" I asked, noticing how tired she looked, and also sad.

"I am missing my children and husband so very much. They are my life, my love and I feel today I have been gone too long." Her eyes then lit up, "But I have had word that in one week my husband will come to join me to walk the rest of the way."

"That is fantastic news," I replied, "I know your love must be very special for his ongoing support of your need to be on this journey."

"Oh Tess it is, we are each other's best friend. I have been there for him and now he is here for me. We grow together in union by encouraging each other to fulfill our soul's purpose. Sometimes it takes us on journeys, sometimes it is just holding each other tight."

"I can so appreciate that, and hope a day will come I can be with someone that can share the same."

We hugged each other in support. I secretly hoped we would see each other again, but said goodbye, not knowing.

After my wine break, back on Nelson, my quads had begun to ache terribly on those last kilometers to Leon. The trail consisted of small rolling hills, and I was certain the translation of "mostly" from Spanish to English was very different, I chuckled to myself. I had been pushing Nelson up and hanging on going down more times than I had expected. The rolling hills that day allowed me to stay peddling, but it was a serious strain on my quads. On the brochure Oscar had given me about cycling the Camino, a whole page was dedicated to proper training. Needless to say, I had not spent time working towards 45 kilometers a day, nor had I practiced handling a bike fully loaded over rocky terrain. It was useless information, but enough to give me a laugh as I had given up on "mostly flat" because I really no longer knew what it meant.

On the last 10 kilometers before Leon, I was presented with a brand new set of challenges. No longer in the forest but on busy streets packed with motorized vehicles, I biked through the outer edge of the large city. I came upon a narrow bridge with no allocated pedestrian sidewalk and had to pedal like a scared rabbit with a massive truck on my back wheel and cars approaching the other way at lightning speed. I was having a little heart attack for fear of becoming Spanish roadkill! The landscape had turned industrial, the homes lost charm and history, and there were massive B train semi-trucks and so many people everywhere. I had not expected this on the Camino, but for a path that was carved out 1000 years ago, the world had grown up around the Way. To any pilgrim at that point, it felt invasive on the senses but understandable. I stayed focused on getting to Leon, and the end of my 60 kilometer day.

A mere five kilometers before the core of Leon city, I looked ahead to what lay before Nelson and me, a very distinct up, there was no questioning the dread in my mind. I pulled over for a moment in the quiet suburb I was now in and plunked myself down on the curb to have a smoke and contemplate the stupid dirt hill a kilometer away. No matter how hard I wished it to miraculously dissolve, it wouldn't. I hoped and prayed this would be the last one as I shook off the dread and leaped almost too enthusiastically on Nelson and then it happened! My saddlebags, boots, and Harold fell off, taking Nelson and me to the ground as well. Lying in the dirt with a skinned elbow, I sat in the middle of the chaos and laughed so hard I almost peed. After all, I had been through the lesson of learning to laugh at myself and it came easily. As my Korean friend had said, "What else can you do?"

Regrouping myself and my array of things, I clambered back on Nelson only to discover my handlebars did not look right. My first thought was I'd bent the bars in the crash so I tried to push them back up, nope that didn't work. I then realized all they had done was spin around backward and I was relieved that with an easy fix of basic female engineering, I was back on my way.

LEON AND CONNIE

Arriving on level ground, I was instantly in the city. The path dropped me into a major intersection, which abruptly took me out of my dream state to look for the yellow arrows to guide me into old Leon. I had to deal with sidewalks, pedestrians, traffic lights and fast-moving cars, and pass large modern office buildings, high-end clothing stores, and fine dining restaurants until I finally arrived in the part of the city that stood the test of time. Leaving the modern and coming into the old was where I felt a deep connection with all the beauty Spain stood for. I rode Nelson down narrow streets with the fresh fruit stands, small bars and bistros spilling out onto the sidewalk. As I rounded the last corner, the arrows brought me to the center square where I was unprepared for the magnitude of the gothic cathedral that stood before me. Parking Nelson, I sat on one of the many benches to inhale the beauty of the ancient structure. The towers of each corner stretched up so mighty and strong they almost disappeared into the heavens. The cathedral itself appeared to be elevated in its entirety into the sky yet rooted in its earthly connection. The gothic detail surrounding each arc of the windows told stories of mystery and belief. The massive wood doors at the entrance, hinged with black steel, could not have been less than forty feet tall, protecting its interior yet holding out a hand to those who would enter. The square was alive with an energy I could feel in all my senses. The children in their school uniforms were playing, dancing and laughing under the watchful presence of the cathedral. There were lovers kissing on the benches, old women with their purchases catching up on family gossip with friends, men playing chess, families and friends everywhere engaging with one another. Like I had observed in Pamplona, no one had their heads buried in technology, instead, they were talking and laughing face to face.

114

The sun had begun its descent, so I got back on Nelson to go in search of a place to stay for the night. I biked up and down streets around the square looking for the familiar symbol of a hostel but strangely could not find one. I did discover a quaint hotel not far from the square and, with no other option, I checked into La Petite Leon. I inquired about storing Nelson until I could make arrangements to have him returned.

"Who did you rent from?" the desk clerk asked me.
"Oscar in Burgos," I replied.
"Ah, we know Oscar," she said. "I can arrange for his man to come and get the bike. One moment please."

The clerk made a brief phone call while I filled in the paperwork and, mere minutes later, an older gentleman showed up and unceremoniously walked off with Nelson. There wasn't even time for a goodbye to my beloved comrade who bravely accompanied me over the 180 km. I knew I was being silly but I had formed an attachment to Nelson and quickly caught a photo as he rolled away. Another Camino lesson: physical attachments to things are not necessary for happiness, it was the memories I would treasure. After dropping the purple monster off in my room, I sat outside in one of the many cafes surrounding the holy structure. I savored an ice cold *vino blanco* as I watched the sun setting behind the Cathedral in awe. It grew in height as the last rays of light filtered through the carvings adorning the walls with beads of color bouncing off the stained glass windows.

"Tess, is that you?" a familiar voice called out. Looking up, I smiled. It was Victor, the veterinarian from Estrella.
"Victor! Hello, wow so great to see you again." I stood to embrace his outstretched arms.
"I am on my way to have dinner with some great people I have met. Please come with me, we can catch up as we walk."

Without a thought, I downed my wine and fell into step with him. We exchanged our experiences since we last saw each other. He was astounded how much I had gone through and commended me for my will power, not before wagging a finger at my stubbornness.

We arrived at the designated restaurant and approached the outdoor table where at least twelve other pilgrims already sat. I was instantly excited to have some social time with others from the Camino. Victor did his best to introduce me to everyone and I did my best to remember names. The table was alive with conversation, laughter, delicious plates of tapas, bottles of wine, and Camino stories. I enjoyed sitting back, listening and observing all the different characters around me. A young woman sat beside me and introduced herself as Alex, she was holding hands with Victor. Lovely to witness another Camino love brewing. To my left sat a German man, his name was Hans, with an intense look on his face. There was a mature connection between us, sitting amongst the youth, and we spoke quietly to one another. We discussed how we both had the need to repeatedly check everything in our packs. To go so far as to look in each compartment just to reassure ourselves we had each item in our mental checklist. There was always a feeling of relief when the task was done. Hans stated that there is always a part within us that has fears that we can't let go of. I expressed skepticism so he took his thoughts further.

"When we are young, fears surface based on the unknown due to lack of understanding, knowledge, and experience. As we become adults the same fears follow us until we apply the learned knowledge and reprogram our subconscious through experiences. The thing about our packs, we can rummage through them the instant we fear a loss and by doing so that feeling of fear is promptly gone. Why be afraid of something you can resolve instantly? We all have the ability we just have to be able to rummage, like our packs, through our minds."

"Yes, I understand that and I am experiencing that very thing. I feel I can better address my fears and have let some of them go, yet not all."

"Why do you walk the Camino Tess?" Hans asked.

"Originally I was walking away from an uncontrolled, dissolved life, but then once away, I thought it more of a penance for not being kind enough to who I was." I paused and took a deep breath, "Now I walk with the joy of who I have finally allowed myself to be." Hans patted my hand and smiled at me.

"You, Tess, are truly getting the best of what the Camino can give you. I wish you nothing but fearless acceptance of the joy that will unfold for you." Hans stood and waved goodbye to the table and just like that disappeared into the night. I thought of the words I had expressed and how easily they had come. I too stood and said farewell to all, with a hug for Alex and a wink to Victor, and headed back to my hotel through the labyrinth of streets. Feeling a little tipsy from the copious amounts of wine shared, I walked a little further than needed and ended up in the cathedral square. I nestled into a cafe chair and ordered a warm coffee and a ridiculously large pastry, life at that moment was right on track.

Waking up that night at 3 am to a boisterous party outside my window, I could not fall back to sleep, so some Facebook time it would be. A message from my friend Heather brought tears to my eyes. Connie, a wonderful vivacious woman who could radiate light on the darkest days, had been diagnosed with cancer. The dreaded disease had taken residency within her adorable, freckled pixie nose in her sinus track. I was so sad and angry as to why the fuck this could happen to such a beautiful and kind person. Connie had already been a proven survivor of personal matters in her life—now this? The whys in life that cannot be answered are the hardest to let go, though we still crave it, we demand it, we want it and the answers still do not come. Spending time with the questions that cannot be answered only serves to prevent your thoughts from moving forward. I would go to Leon's majestic cathedral and pray for the first time in a long time for my friend Connie. I believed I would be heard. With love and light, I sent her a message from Spain:

"Sweet Connie...I don't understand...you are an incredible soul. My words from across the world come with the conviction for your healing to begin right now! I will walk with you in my heart my dear friend, I will pray for you in the cathedrals that have stood the test of time. I will believe angels like you are needed on this earth. My love, my gratitude dor being your friend are coming in waves upon waves upon waves."

I laid back and looked at the stars through the window in my slanted ceiling and thought how life can change so quickly and without even knowing so. I became aware clearly at that moment that we must command ourselves to live to our highest potential. To treasure all the moments and to experience the opportunities that cross the paths that excite us on any part of life's journey. We must celebrate our body's ability to function, to love it and to care for it. To embrace the love that comes from any direction and pay it forward in return. As I closed my eyes I had a beautiful vision of my three children smiling at me.

THE GAZELLE

Great news arrived via email the next morning, Carlota and Jorge were on the bus from Sahagun and would be arriving in Leon soon! Funny enough, Carlota had also booked the two of them at La Petite Leon, unknown that is where I was staying—ah the Camino Way. I sipped on my morning coffee outside the hotel in anticipation of their arrival.

I thought about my friend Rene who had completed the Camino the previous spring and how one person's Camino can be so different from another's. Rene had shared with me, in our many my pre-Camino chats, that in truth, he walked in hopes of finding love. He had relished in his daily routine, chatting with everyone, partaking in nightly *albergue* gatherings, and giving out his free hugs to anyone who wanted or needed one.

It was then that my family arrived and we were all so happy to be reunited again. Carlota and Jorge quickly checked in, put their packs in the room and the three of us strode arm and arm into the square. Carlota and I talked nonstop while dear Jorge just smiled and nodded, trying to keep up. She and I felt a kinship of sisterhood neither of us ever had before. A small dream coming true for both of us.

Coming into the square, we all agreed we could not wait and went straight to the doors of the Cathedral Leon to explore inside. The moment I crossed over the threshold, I could physically feel the powerful energy through my sandaled feet radiating up my legs and into my spirit. My eyes went straight to the large stained-glass windows everywhere possible, sculpted into the high stone walls. The greatness was nothing I could have ever imagined with the intricacy and details of individual stories told on each pane. Even though it was built a thousand years ago, there was no way modern engineering could build such a powerful structure, not just in materials but also in the strength of spirit. There really had to be a belief and connection to a God-like spirit, for what else would mentally empower men from so long ago to construct it. Jorge explained to me all the churches of Spain built the altars facing east so that the rising sun would shine upon them. He described that Cathedral Leon was unique in that its magnificent stained glass windows were designed around how the sun's light cast upon them. The window on the right of the altar was in shades of yellow, orange and red, which would absorb the mid-morning rays and on the left, shades of blue, grey and black captured the setting sun's glow. It is there I knelt to pray for Connie to stay amongst us, for angels like her were needed on the earth. I felt a movement and Carlota knelt beside me, strengthening my prayer. It was a moment I would never forget, I felt God that day.

Departing from the cathedral, quietly lost in our own thoughts, we were magnetically drawn to a lively cafe. Jorge broke the silence in his jovial voice, "Come, my girls, let us eat!"

For three hours we ate, drank and roared with laughter as we always seemed to do when reunited. Jorge had some marvelous stories to share from their days of walking. He began with a recount of their passing across the 17 kilometers of the Mesta when all was silent around them.

"There was a professional cyclist—you know the kind in loud stretchy tights—coming up behind us," Jorge began. "As he rode past us he shouted, 'Not too far, keep going.' Carlota and I said to each other he must have GPS or something and we both felt invigorated to increase our pace if it wasn't too far. Well, little sister, two hours later after we both said nothing to each other because we were pondering his words, Carlota broke the silence and said, 'I think his GPS might be wrong.' That was it for me, I lost it. I could not believe this biker could basically tell us almost there! There was NO almost there! We were never almost there!" I was laughing so hard as Jorge told the story with his arms flying about reliving the absolute honest emotion he had felt. Not yet finished with his rant, Jorge continued, "If I ever see him again I will punch him, his dog and put holes in his tires." Carlota and I had tears rolling down our faces as we doubled over knowing of all people, Jorge would never hurt anyone or anything. I could understand his frustration, remembering "mostly flat" and pitied the man in the stretchy pants if Jorge ever found him.

Carlota then told a story of a young tall Columbian man they had met on the way out of visiting a village church.

"Tess, he had asked if he could walk with us for a while to help the time go by faster and, of course, I said yes. We were quick to learn this young man was covering on foot 40 km a day and being over six feet in height and mostly legs it gave him quite a long stride," said Carlota. I immediately had a vision of my two small-statured Spaniards almost running to keep up the pace of the Columbian.

"But he requested to walk with YOU."

"Yes, yes, but we would be damned if we didn't keep up with him," Carlota replied. Jorge now snickering added, "At one point I looked back at my little love and asked how her feet were doing? She gritted her teeth and replied, 'they are fine, just FINE.'" Carlota finished the story explaining at the next village, they said polite goodbyes to the Columbian gazelle and collapsed at a cafe in exhaustion. I could so see the two of them laughing at each other for what they had just accomplished, keeping their pride.

"So is it correct to say the Camino is bringing you closer together?" I asked.

Looking at each other and clasping hands before looking back at me they both replied in unison, "Yes."

I smiled at the answer I already knew, as I watched them together and could see the bond growing in all the little things they were sharing. Jorge was a kind and gentle man engaging with Carlota who though tiny in stature was so strong in mind. The times I had spent with her, I had experienced her uncanny ability to organize like a corporate executive. When things were needed or a place to stay she could find it in no time.

After lunch, we went back to the hotel and assembled our clothes and set off to a laundromat. To give our meager wardrobe, that had only been hand washed until now, a good cleaning. Carlota and I left Jorge in just a pair of shorts to supervise as we snuck off for a coffee together. We felt blessed to have that time alone as we discussed women of different cultures with a comparison of our own lives. It was empowering and brought our relationship closer in that stolen moment. An hour later, gathering up Jorge and our freshly cleaned laundry, we headed back to the hotel. We said our goodnights early, I was desperate to journal and they were ready for sleep.

I HEARD MY VOICE

"She dances to the songs in her head, speaks with the rhythm of her heart and loves from the depths of her soul." - Dean Jackson.

That very statement resonated through me when I woke up the following day. I was going to start walking my beloved Camino again, and now, with Carlota and Jorge. My blister was almost closed over, the hip no longer an issue, and Carlota's feet were attempting to feel better. It was Jorge who suggested the idea over morning coffee, his love for us first and foremost:

"Let's take a taxi through the industrial outskirts of Leon. After the beauty and peace we have enjoyed here, let us begin the Camino in nature." Both Carlota and I wanted to argue that we should walk but with the stern look on Jorge's face, we thought better of it. Jorge, who was always so easy going, and went along with anything we wanted, had a good plan. As we flew past the 10 kilometers of the not so pretty parts, I no longer felt guilt or failure for not walking everything. Instead, I understood the Camino was a journey of so much more than the kilometers walked.

We were deposited on the side of the road where the Way leaves the pavement and disappears into the bush. Crossing the road, we stood in front of a small winery that was already opened to the public.
"Come on let's go inside for a sample," Carlota said with a wink. Giggling at the fact it was only 10 am and we had just taxied, we felt like naughty school children. We took turns sipping different types of delicious chilled red wines and after settling on a flavor we liked, sat outside under the umbrellas.

"I love this, I love that we can be free enough to enjoy such a simple luxury," I said.

"This little sister is what life is all about. No rules," Jorge replied, gazing at Carlota.

I marveled at the love that was obvious between my two friends. I had come on this journey with a shattered heart, yet those two souls, along with others, were showing me in bits and pieces what real love could be, not the illusion. Love surrounded me every day and I was seeing it through new eyes. I silently lifted my glass to the Universe and toasted with thanks for putting me there, right where I needed to be.

The remaining 16 kilometers of the day were walked in peaceful, easy silence, from the glow of the morning wine and the comfort we felt together. Sometimes we walked side by side and other times we drifted apart but always near enough to be together. We arrived in the village of Villar de Mazarife where we stopped at a small cafe for cold drinks, the day had become very warm. Carlota, our clever, masterminded companion mentioned she remembered not far off the Way was a watermill house that took in pilgrims overnight. It was agreed it would be a fun adventure and Carlota went off with her phone to make the arrangements.

"Are you enjoying with us Tess, doing the Camino sometimes, the 'other way?'" Jorge asked when Carlota was gone.

"Oh yes very much. I have seen more, slept more and saved my feet for better trails. It has taught me to live under no one's expectations, only my own and I am no longer so hard on myself because of it."

"I too have always been trying to please others before myself and it got me stuck in a life I didn't enjoy for many years. Now I look into my desires and follow where it takes me."

"Like loving Carlota?" I asked of him.

"Yes, she is becoming the light to complete me, for with her I am feeling a love beyond what I could have thought possible. Doing the Camino together is like living a lifetime, with all its joys and struggles and growing through them together."

I stood and hugged my new brother for an admiration I had never felt before in a man.

Carlota then returned to us from her phone call very pleased to tell us the owner of the house would be by shortly to pick us up. The house was six km off the Camino and Carlota had made an executive decision to accept the ride on behalf of our feet. Jorge and I smiled in agreeance.

Driving down the long solitary driveway, there stood a simple yet lovely stone millhouse nestled amongst the trees. Carlota told me its history dated back to the 13th century, and, seeing it, I felt a lump in my throat of a place I had been before. Clambering out of the van, dragging the purple monster, I stood with my face to the sun and inhaled a deep breath of autumn air. Carlota and Jorge followed Marta, the owner into the house, while I sat in the grass. I undid my sandals, peeled off my socks and all the bandages that encased my battered feet, and stood to walk circles in the lush green grass. I was determined it was Mother Nature's turn to heal my body in a place I felt at home for the first time in a long time. I would have walked a million kilometers to be in that astounding moment full of pure natural spirit and energy. I heard the familiar voice of my sister calling out to me, "Tess come in, there is so much for you to see," Carlota called out from the doorstep.

"I believe you, I am coming." I scooped up my discarded items and pack and made my way to the house. I stepped inside and drank in the beauty of the interior, the openness drew me to the center of the house. Outside the French doors on the left side of the great room, I could see the stream of water coming out from under the home that once would have powered the mill. The feature in the large room was a massive stone fireplace that stretched from the hearth to the high-reaching ceiling. Two large overstuffed chairs graced either side of the warm fire with an invitation to curl up and appreciate. A staircase along another wall was built with beam wood steps and the handrail was carved in delicate whimsical designs. In the center of the room stood the beating heart of the house, a long rustic wood table with a dozen mismatched chairs. It was there I could feel the familiarity of those who had and shared stories, tears, and laughter.

Marta, our hostess of the magnificent house, had disappeared into the kitchen and in her place emerged Max, her husband. The instant he walked into the room he emitted a joy that all three of us felt.

"*Hola, Bienvenidos peregrinos, cansados a nuestra casa!*" With open arms he hugged each and every one of us. Carlota whispered to me that he had expressed great happiness we were there and sincerely welcomed us to their home. I was understanding more of the Spanish language, especially with the aid of facial expressions and hands, though I was grateful to Carlota when she translated the specifics. As Max continued, conversing in rapid Spanish with Carlota and Jorge, I watched and admired his vibrancy and passion. He was tall like an old tree with snow white hair and a greying beard, dressed in traditional cotton whites that flickered faintly in the natural breeze of the house. Max poured homemade wine from a large antique sideboard under the stairs and passed the full glasses around to toast our arrival. Marta came from the kitchen bringing plates of meat, cheese, fruit, and olives from their garden. We all gathered at one end of the stoic table to take our seats, noting we were the only guests for the day. The joyous banter amongst the Spaniards in their native tongue had begun and I listened more mesmerized by the beauty and passion of their voices than what was being said. At the same time, I was taking in all the details of my surroundings. The Dutch wooden shoes in the corner, baskets full of well-read magazines, walnuts drying on the mantel and the eclectic display of mementos that all spoke their own stories. I felt I could live in that place forever.

Breaking me out of my daydream, Carlota tugged at my shirt to tell me Marta would be taking us to our rooms. Together we were led up the solid staircase and shown to our respective private sanctuaries. Jorge and Carlota were going to have a nap under their hand-carved tree of life headboard, and my room, with its quaint balcony overlooking the yard, was perfect. Marta informed us that dinner would be at 9 pm and asked if we would like to join them. Carlota was quick to respond on behalf of us with an enthusiastic, "Yes."

Unloading my pack, I found my way back to the garden with my journal and laid out in an inviting lounge chair in the setting sun. It was not long before I had the company of a tiny full-grown cat who nestled in a ball on my lap. The serenity and peacefulness took me to a wondrous place of rest.

At 9 pm, we gathered again at the table that was laid out with a decadent feast, not only for the appetite but also for the eyes. The array of delectable treats that I soon would be savoring was more than I could believe.

Carlota came up behind me, "Looks good, yes?" she asked.

"Oh yes, I am in awe of the food, the house, Max, and Marta," I replied. "It is all so perfect."

"This place is very special for me tonight, I was here the last time I walked the Camino. I have brought us here to create a new beautiful memory." I saw a look in her eyes I had seen before when she spoke of her previous Camino. Not certain exactly where her thoughts went, but if being here again was as good for her as it was for me, a beautiful memory it would be.

We feasted on fresh zucchini soup, warm homestyle bread, succulent tomatoes, plump juicy olives, pork basted in watermelon, red wine and finished with Max's homemade liqueur. The meal was not only for the body but also for our souls. After dinner was finished, we continued to sit at the table as I listened in earnest to the four of them engaged in a detailed conversation. Emotion in any language is universal, and even though I did not understand everything being said, I felt a part of it all. When a funny story was told I could not help myself but burst out laughing with them. Sweet Carlota would interpret just enough words when she could to help me. I glanced at the clock on the wall above the sideboard, I could hardly believe it was after midnight and I delighted in the long lost art of dining table enjoyment.

Through Carlota interpreting, I learned Max and Marta had been married for fifty years, together since their early teens. They boasted of their twin daughters who have each given them a grandchild. They also shared they were going to sell the millhouse and move to Madrid to be closer to their family, it was time. Carlota was quick to inform the couple that three of us would buy the house. I fell in love with the thought and Jorge rolled his eyes so tenderly at his little dreamer. I saw in Max and Marta, after fifty years of love, what I also saw blossoming in Jorge and Carlota. I believed that one day, I too would have the same love I was witnessing with the two couples. The Camino was showing me the truth that comes from a pure love that is unconditional.

"Do you know Tess can play guitar?" Jorge announced in Spanish to Max. Carlota was quick to translate to me because she knew what was coming next as the conversation of music and talent had led up to this announcement. Max was gone in a flash and I started to panic and when he returned with a very old nylon string guitar. I whispered to Carlota,
"Ah shit, I can't play in front of people, please save me."

She just smiled at me as the instrument was placed in my arms by Max with words of encouragement from all. My fears started to seep away, I was not the same person anymore and holding that guitar in that amazing warm atmosphere amongst unconditional love, the choice was mine. Hearing a whisper in my ear to play, just let go and play, I began my favorite song about a bird flying into the horizon. I got quickly lost in the music I was creating and much to my surprise I started singing the beautiful lyrics, hearing my voice carry through the perfect acoustical room. A voice that was coming from a place I had never sung from before carrying so much emotion. When I finished, a happy tear rolled down my cheek knowing I had achieved something so very personal. Both Carlota and Jorge stood up and came to hug me, as they both knew enough to know what that moment had meant to me. Max and Marta applauded with enthusiasm. I handed the guitar back to Max with thanks, thinking about how I had found a connection to myself. I felt a glow like never before from deep inside and understood it. I would lay down the walls that had been preventing me from being the true me. With Leonard Cohen's Hallelujah playing softly in the background, I excused myself from the table and curled up in one of the armchairs. My friends carried on in their own soft conversation as I settled into the warmth of the fire and my thoughts.

I heard my voice from within, it was calling, not what I had expected ever but the clarity was surreal. I would write words that could be shared with others. Unsure how to nurture it, only knowing, without doubt, I could and would trust my journey. Society needed to be reminded to live life through adventures and experiences, whether around the world or around the block, there are no excuses. I knew I had, especially in the last few years, got caught up in what I thought life should be rather than truly experiencing it. Our wonderful lives have been designed for us already, but it does not come unless we choose to live it, believe in the possibilities and keep the faith. I struggled for years to find the answers from other people, seeking resources to guide me, everything outside myself to show me how to find happiness. I meditated kind of, yoga half-assed, journaled off and on, walked though not far, listened to music to cry and self-help books that claimed they would show me my happiness. None of these had given me the complete answer and I had been frustrated with money and time ill spent. Why did all the gurus "get it" while I remained in the dark? Until that moment, in that beautiful house in Spain singing from my soul did I finally understand. Possibly all the things I had ventured through to get here was a prelude for the experience. I lived in a state of awareness that there was so much more to life. I had been a dedicated seeker and was becoming a finder. I knew how lucky I was to have this time on the Camino to find answers but I had no choice. I could no longer carry on the way I was trying to live my life, it was not working. The answer I needed had always been within me, it was to love myself unconditionally. We are all born seekers to find the life we must live. I smiled inside knowing I had found my purpose.

THE UNEXPLAINED

Starting day twenty-four, I was feeling more alive than ever. I had slept for eight hours under the comfort of the homemade quilts and warmth of the house I felt secure within. My energy had shifted to an earthbound pull for the Camino, no longer a conscious walk it was just what I had to do. Excited and eager to start, I packed and bounded down the stairs to be greeted by my family eating breakfast.

"Good morning little sister, aren't you a ball of energy with glowing eyes," Jorge said.
"Oh I am, I am. I feel so good and excited about my life today!" Carlota stood up and gave me a big hug and whispered in my ear, "This is your *Beun Camino* Tess."

Max and Marta came from the kitchen with cafe con leches and joined us at the table. Max had a Camino story he wanted to share with us before we left and specifically told Carlota he would speak slowly enough for her to be able to translate for me.

"A small village not far from here, a pilgrim had a strange experience. He had been walking all day and, of course, was tired and sore, but more so overwhelmed with many emotions. He sat outside a building on the curb crying with his head in his hands. An ancient Spanish man approached him and lay a hand on his shoulder in comfort and asked if he was alright. The man looked up into the wise face and told him no, he was feeling far too much. The old man suggested he come with him to his home at no cost, he would feed him and give me a warm bed for the night.

In the morning the pilgrim, feeling rested and full of peace, searched out the kind man to thank him for his kindness but he was nowhere to be found. Bewildered, he wandered around the village looking until he started asking others if they had seen him. No one knew who he was talking about. There was no one in the village that fit the description. They suggested he meant the young boy who ran the *alberque*, but he was away in another village. The pilgrim insisted the man he was trying to find was very old and still no one could help. You see the Camino brings you magic when desperately needed, to help you believe the answers you are receiving come from within and reward you with peace." Max finished his story, stood up and clapped his hands together. "Now my pilgrims, go find your magic!"

Leaving the millhouse along a trail that would rejoin the Camino, I thought of the story Max had told us, which was strangely familiar to an experience I had.

"I must share something with you two," I said to Jorge and Carlota. "When I was five days on the Camino, I was walking very much alone, with no one in front or behind me. I was listening to music, singing along and observing my footfalls over the rocks. Something made me look up and not three feet away from me was a man dressed as a gypsy. He wore short baggy pants held up by a rope, old beaten boots, a cropped black coat, a red bandana around his neck, and a black beret. He was leading a scruffy donkey with an ancient rope attached to the makeshift halter. The donkey was packing clothing, tools and a large bundle of cut grain. The man and donkey were coming from the opposite direction towards me and we made instant eye contact. He asked in a unique dialect, that I interpreted, could I spare five pesetas to feed his donkey? I reached into my pocket without a thought and handed him a 10 euro coin. He clasped his hands together in joy, smiled and was off again with the donkey in tow. I had not given that moment much thought until after hearing Max's story."

"Tess, no one has walked the Camino in that manner since the 1940s, asking for peseta not euro, well that currency dates back to the 1800s, and going the opposite way…?" Jorge said.

We all thought for a moment then Carlota said only one word, "Magic."

After five kilometers, we stopped for a warm coffee and to layer up as the day was cooler than we had thought. The sun was out providing beautiful light but no longer much warmth. Jorge and I got into a discussion about television news and we discovered we had both given it up eight months prior. Jorge, like myself, found too much sadness there and it would only serve to torture our empathetic souls as there was so little we could do. The constant badgering of negative information, mostly dramatized, suppressed a mindset to what we could do in our immediate surroundings. Carlota reminded us of a quote from one of the most empathetic icons, Gandhi, "Do for yourself first what you want for the world." We all agreed that the constant barrage of doom and gloom was only brainwashing us to believe there was no hope. Living in a society where men, women and children can eat their meals while visions of horrific proportions are displayed on the screen, which only resulted in immunity to the destruction of humanity. Also, discussing the saving of another part of the world made no sense when not enough is done for those that live in our own neighborhood. We concluded the topic with, as individuals, we should abide by Gandhi's quote and be the best people we can be for everyone else to see.

After the rest stop, we decided to part ways for the day. I was ready for some alone time and Carlota had a place she needed to revisit from her last Camino, with Jorge. Her eyes showed me the same look I had seen twice before and I then understood more of Carlota's quest on this Camino. The energy of love between us as we hugged goodbye was very sincere, an intensity I could not explain. I hugged my Camino sister with all I had and wished her new beginnings in this place of hers.

The dirt road I traveled on was dry and dusty and I could picture the nomads following that exact same road centuries ago with donkeys and carts. There were no other pilgrims anywhere, I was enjoying the serenity that comes with silence and solitude. I walked up and down rolling hills, feeling the joy of the previous night's revelation and I was grateful my wound was no longer an issue. I was letting go of the worry of time, the kilometers and reaching Santiago, for in my heart I knew it would all be okay. Knowing Carlota and Jorge would eventually travel the same path, I drew a large heart in the dirt with their names, thinking to myself how sweet it would be if they found it.

From there, the Way returned to the familiar steep mountain climbs that I had not endured for some time. I reached down to regularly adjust the Velcro on my sandals to avoid blisters and pressure points. I had tried to put my boots on every day but the new extremely sensitive skin would not allow any pressure, so the boots remained slung across my pack. After cresting a large hill, the descent was extremely rocky and I would need to focus on negotiating it with care. I pulled Harold out of hibernation from my pack and lengthened him, a wonderful action he could do, that would keep me balanced against the downward pull. Unfortunately, something inside Harold broke and he could no longer be shortened, regardless he faithfully assisted me to the bottom of the hill and returned to my back in his permanently lengthened glory. I thought with a smirk, maybe a flag might be suitable, as he then protruded fairly high above my head.

I came upon a very magical place in the desolation of that part of the

Camino. Appearing out of nowhere were the ruins of a stone house with only three walls remaining and a garden that was a mystical oasis. First was a small labyrinth made of stones and, knowing the ritual from my Wiccan practices, I proceeded to walk the maze repeating my own affirmations.

"I will always love who I am, I will trust in life's journey, and I will believe in my inner and outer beauty."

I was able to repeat this many times before reaching the center of the spiral where I dropped my token. The next discovery was an alcove, that may have once been a room of the decaying house, and in it was a mattress with a ton of pillows. I climbed in to enjoy the comfort and to my delight, I was welcomed by a momma cat and her brood of seven kittens. I relished in the opportunity to snuggle and love each and every one of them and stroked momma's head for how good she was. The oasis continued into a rose garden showing years of unkept growth with a cart set up with fruits, juices, bread, and jam. It was all so perfect and I gratefully reached for a large red apple.

Somewhere from behind the walls a man appeared and in English said, "Hello, hello, please take all that you need. The walk ahead is straight and long."

"Thank you, this is wonderful," I said as I reached into my pack for euros.

"No need to pay for my gift of love, as it should, it is free for you." once again the Camino provided. The man stamped my rapidly filling up pilgrim passport and I knew it would always be my favorite, the one with just the simple red heart.

Beginning the long, straight stretch of walking ahead, I munched on the apple and I thought of Carlota and the side journey she was sharing with Jorge. I felt certain she was releasing something from her past there and having Jorge with her signified the importance of going through not only the good but also the not-so-good of our lives with the someone we love unconditionally. I knew without a doubt these two would make the entire life journey together.

The kilometers were proving very flat, hard and lonesome. I did not listen to music, instead the silence as I put one foot in front of the other. An ache developed in my feet as well as my back and shoulders, I was trying to think of ways to lighten my pack. As I sorted through what I could easily discard in my mind, in the distance I saw a stunning concrete cross in the setting sun. Moving closer, it grew in height and I approached with curiosity. Looking past the cross, the descent began and the city of Astorga was in view. What lay beyond was more breathtaking, the magnificent mountains of the Galicia's, which I would soon cross. Carlota had told me that morning, I would see the mountains of her home region but she could not have prepared me for the vast magnitude. The realization of the distance I would walk to cross those mountains was nothing I could begin to comprehend. Those are the mountains that would take me to the place I had been anticipating since I had left home, The Cruz. The stones I had carried in my pouch for over 7,000 kilometers from home would be placed there. It was a pilgrim tradition to place the stones brought from your homeland at the base of The Cruz, to symbolize leaving your sorrows and to begin again. I had four.

Arriving at Astorga, first finding a hostel, then dropping my pack, I set out to meet with Carlota and Jorge as they would be arriving soon. Grabbing a quick bite to eat and exploring the smaller city I saw them coming into the square. It was so good to see each other again, though it only been half a day. I noted that Carlota looked a little different as she stood hand-in-hand with Jorge, peace had settled in her eyes.

"Did you accomplish what you needed to do?" I asked her.

"I did, I have released a memory that was weighing me down. To have Jorge with me for support, I am now free to completely love him."

"Aw, my sister, I am proud of you and your fearlessness and Jorge," I turn to him, "You are a man of honor."

" I love this woman Tess, it was nothing." he wrapped his arms around Carlota's waist and planted a kiss on her cheek.

"Tonight we celebrate with each other romantic time," Jorge said as he winked at me.

"Love you both for who you are becoming, together. We shall see each other in the morning." I hugged them feeling their love and wandered off into the square.

DON'T QUESTION

The sun had not yet risen in Astorga as I sat outside with my cafe con leche and the purple monster the next day. The cathedral would be open in an hour, and I was going to meet Carlota and Jorge there. I had tried again that morning to put my boots on, but they still aggravated my sensitive skin. So once again they were strapped across my pack, along for the ride.

I gazed up at the Galicia mountains that we would walk that day, intimidated by the steady climb but inspired by The Cruz that awaited me at the top. I had begun to feel the transition to the third stage of this journey, the spirit. With only 280 kilometers to go until Santiago, a sadness was lurking within me. It had been sneaking up since Leon. The pilgrimage had become a way of life and I could not imagine what it would be like to be back home sleeping in the same bed every night, a closet full of clothes, and driving my truck. To continue the lessons learned was easy when you lived and breathed them on the Camino, but to take the lessons home out of the pilgrim's lifestyle was frightening. Letting those thoughts drift away, because it was too much to think about, I paid my bill and headed to the cathedral.

" Good-morning my family!" Smiles and hugs awaited me when I found the two people who meant so much to my heart.
"To you too, little sister. Did you sleep well?" asked Jorge.
"I did. I feel super rested and ready to take on the day's climb."
"The beauty of the mountains I hold so dear to my heart, we will feel so much," said Carlota.
"And for that, we need to eat, let us go have breakfast before we go see the cathedral," Jorge said.

We found a pleasant indoor bistro with comfortable chairs and ordered fruit, bagels, yogurt, and juice. I took the moment to show Jorge poor broken Harold with hopes he might be able to fix him. Jorge took this task

132

very seriously and examined the cane carefully, for he too thought highly of it.

"Let me see... if I turn this... and pull here…" A sharp snap echoed in our ears of Harold became two parts. Jorge held a piece in each hand and for a moment there was silence. It was Carlota and me who broke out in laughter followed quickly by Jorge

"Shit, I am sorry." Still laughing, "Well now he has a twin."

"Jorge you crack me up, not to worry it is all good. Harold has served his purpose." I said.

I took the two pieces from Jorge and strap the twins to my pack, not ready to give them up yet. As we feasted on breakfast Jorge told me the story of their romantic evening with the flare and humor I had come to love.

"So," he began, "After we checked in, we went downstairs to have a romantic dinner in the hotel restaurant. We were set near the door with a definite draft in the very empty room. I asked, 'Can we have another table,' you know me, Tess, being romantic." Chuckling Jorge continued, "The server says there are no tables available, they are all booked up. I say to her I have seen pictures on the wall of a much cozier dining-room and she tells me, 'Yes, we have it, but it is not heated and you would be alone.' I waved my hand around the room and said, 'Well it seems we are alone and it is cold here.' I am smiling at her and Carlota, as we think she is making a joke, but she is not. We tell her we will go somewhere else then, she says to wait 10 minutes, I will turn the heat on upstairs for you. I am thinking, okay this is good now and Carlota and I laugh at the situation. She comes back 20 minutes later because, of course, we waited because we are dumb pilgrims." We all had to giggle over Jorge's statement. "So we follow her upstairs through big doors and walk into a room full of elegantly dressed people in our 'clean' pilgrim attire. Now I look at Carlota and without words question what alone must be. We both smiled as we were seated at our table, which would begin the perfect romantic dinner."

"I can't believe it, what a crazy situation," I said.

"Oh, But there is more. I ordered wine and grilled fish. The wine comes and it's fantastic, and then our plates are placed with a butter-fried fish. I am quickly telling the server there is a mistake we had ordered the grilled fish. He tells me, 'Yes this is grilled fish.' I say, 'No, this is fried fish and we wanted grilled fish.' He must think I am stupid because he repeats, 'This is grilled fish.' I throw my hands in the air and say, 'Come on, this is not grilled fish, look it has no lines from the grill on it.' He replied back, 'Oh we don't have fish like that.' Now I'm starting to be a little crazy because Tess I know how to grill a fish."

Carlota and I were laughing so hard all we could do was nod in agreement. Jorge was exasperated telling me the story as much, as I was sure he was with the server.

He continued, "I look at Carlota, who is no help because I can see her trying hard not to laugh, and I try again with the server. The man completely confused takes away our plates. I say to Carlota, 'We have no idea what we will get when he comes back, do we?' We are both laughing at how crazy this romantic night of ours is, and, when the plates come back, they are put down in front of us and what do we have? Exactly the same buttered fried fish... with a nonchalant explanation from the server, I have checked with the chef and he confirms the fish is grilled. We do the fish in the oven because the oven has a grill setting."

We are laughing uncontrollably, it was a story of such silliness and humor. Carlota between breaths added, " I can still see Jorge's face looking at me like, should I hit him?"

We composed ourselves and left the bistro to explore the cathedral, which was holding a full mass with the congregation singing. I felt the voices floating up and swirling around the large stone building as we entered. Every cathedral I had seen, aside from the altars all facing east, had their own uniqueness. This one was built for voices, voices to be carried into the air. We stayed at the back for the service and just listened and embraced the beautiful music. When the mass was over a kind, older nun approached us and placed in each of our palms a small medallion of the Immaculate Mary to protect us. Carlota added the medallion to my treasured silver Camino shell that hung around my neck.

We walked in the rain and, after 10 kilometers, my socks were wet in my exposed sandals and I momentarily cursed my boots. A little luck came upon me as I crossed the path of an older woman selling hand-peeled walking sticks. She had five and I instinctively reached out for one in particular and tried it out like you would anything, to see if it would fit. Walking, turning and swinging it in rhythm, I gratefully handed the woman a five euro coin. We walked in silence and the gap between us grew farther apart as I unconsciously pulled ahead. This happened with us without much thought, we knew we would find each other again, the Camino said so.

The drizzling rain was there to stay as I climbed higher into the mountain fog, the thickness prevented me from seeing anything farther than 20 feet in front or behind. I heard a familiar sound coming from the mist, a soft nicker that could only come from the creature I loved most. Out of the drizzle, a magical silver-grey gypsy horse appeared. I crossed over to the fence line to stroke the head of the magnificent animal that came to greet me. His eyes were deep black pools of kindness and held my gaze as we spoke a private conversation with one another. He reminded me of every childhood fantasy of the horse I would gallop off into a land of fantasy and enchantment. With one final stroke, I thanked him for the visit and walked away, only to turn back a couple of times until I could see him no more.

The Way came upon a village that appeared to be abandoned and eerily desolate. Walking amongst the old dwellings, down a crooked street, I could not see or feel any sense of existence. My steps slowed as I passed alongside a large and naturally decayed house. I was overcome with a sense of life and a magnetic pull of air blowing not around me but directly through me. I was overcome. The next thing I knew I was seated with my pack on the threshold of the old house with Carlota's arms around me.

"Tess, are you okay?" she softly asked me.

"I think so, not sure what just happened."

"When we came upon you, you looked so far away. Your body was there but you were not, I knew I had to embrace you."

"It feels like I lost a blip of time, it was eerie but without fear, I think I have been here before." Jorge listening, with concern and interest on his face, reached a hand out to help me to my feet.

"Come on, let's go," he said.

Together, we treaded quietly farther into the strange village. There were small signs that people lived there, but no movement. We rounded a corner and in front of us was a small unorganized garden with the village church in its center. The tiny path through the garden drew us off the way, around the stone church, and into a tiny open foyer off its entrance. The door was locked to the church so we took turns peeking through the keyhole that could have only been opened with a large ancient three-prong key. I observed the simplistic interior with minimal detail, compared to the many churches I had seen, yet it felt the most spiritual. We stood together for some time in silence.

It was Jorge who was the first to speak, "There is something in the air here."

Both Carlota and I acknowledged we were feeling the same thing and smiled at each other knowing Jorge was beginning to feel a connection of spirit.

"Look here." Carlota pointed to a small, framed, faded transcript on the wall, "It's the history of the church." Carlota translated to English as best she could from the ancient words used. "It says this church stood for years before it was formally acknowledged, possibly before time began. It is found in an old board game in Spanish history as one of the blocks to land on. It was a game of connecting to the 'other world' and, in the game, landing on this square would be the doorway."

Carlota thought to take a picture of the words so she could translate in more detail later, but as she lifted her phone to the frame, it instantly shut down.

"What! That makes no sense, it was a full battery when I started today." Jorge pulled his phone out. " Here I will get one," he said. As he lifted his phone, it also shut down.

Silence descended as the hairs on our arms stood up, we looked at each other, then back to the frame on the wall. I reached for my archaic phone in a knowing way and was able to capture the photo of the words. Jorge, who before then had been a bit skeptical of my beliefs, looked at me differently and said nothing. The things we had discussed along the Way about spirituality and beliefs seemed to have come full circle for him. Walking out of the foyer into the rain, we saw, standing alone, an old weathered wood cross. In silence, Jorge reached down to pick up a stone from the road and placed it on the cross.

"This is the first time I do this, place a rock," he said.

The custom of placing rocks on various monuments along the Camino had been done for centuries by pilgrims for their own personal reasons, just as Carlota and I had been doing when it moved us. Jorge had not understood this custom... until then.

We said nothing as we continued along the village street. I looked over to my right to see a free-standing frame of a doorway protecting the crumbled ruins of the house behind. On the aged wood beam of the frame was carved in rustic numerical fashion, 1841. I stopped and walked towards it longing to touch and caress the numbers. Not understanding exactly what the numbers signified, they meant something to me and I knew without a doubt that I had been there before. I did not go too deep into thought about it, I just felt an acceptance of what circulated comfortably within me.

Finding Jorge and Carlota inside a fenced garden with a small house at the last place before leaving the village, I also ventured in.

"We can have a cafe con leche here." Carlota was quick to let me know.

"Oh yeah yes that would be nice," I replied, not surprised at all that we found life at the end of the road of the strange little village.

A kind woman came out of the house and invited us to please come in to select our coffees. Following her on to the cozy porch with three small wrought iron tables, we dropped our packs and entered her home. A small counter greeted us with a display of snacks, bits of jewelry, and an espresso machine. Beyond that the living room. Delighted in our find, we ordered and paid for our coffees and were instructed to go be comfortable outside and she would bring them to us. I headed out, leaving Carlota and Jorge to chat with the homeowner. I sat in the garden full of colorful flowers, it was a sharp contrast to the greys and browns since arriving in that Village.

Carlota came out alone and sat with me. "I have a gift for you; a very special token that I know belongs to you." She held out her hand and dangling from her slender fingers on a leather string hung the most stunning Tree of Life pendant. I was speechless as I accepted the gift that spoke a language we both understood. Placing it around my neck and patting it against my chest, I reached out to hug her. I didn't think the friendship could grow any larger than it had, yet at that moment it did, for I had the truth of it on my chest.

Sitting alone in the garden with my coffee and my Marlboro, I left Jorge and Carlota to each other's company at one of the tables. I could not help but watch my two amazing friends engaged with each other. They had their heads together, touching forehead to forehead, legs intertwined under the table and holding hands. It was such a beautiful sight of unity to witness the two of them becoming one. I smiled to see the people I cared for so happy and very much in love with each other. They were sharing a conversation of awakening and an intimacy that only they knew the words.

Leaving the village, we spoke in hushed voices simply because the experience of the village still lingered within us. I pointed out a grove of trees on our right that extended deep into a magical forest. It was an illusion of depth perception, almost trancelike, that beckoned one to enter. The trees casted a protective glow on us as we continued to ascend the mountain. I noticed Jorge looking at the forest and I was certain he was seeing and feeling the same thing. We walked alongside the grove with repeated glances, as though never to forget the sacred trees that possibly stood to protect the village of spirits.

THE CRUZ

Having arrived in Rabanal Del Camino late the night before—after a 26 kilometer climb—I slept well in my loft room. We had eaten at the hostel a dinner of homemade food for the soul that filled our hungry bellies. We were then joined by the young Columbian gazelle that Carlota and Jorge had to jog to keep up with, I was happy to have met him. I left the three of them to catch up and excused myself to go to bed.

The following morning brought a deeper fog and more drizzling rain. I sat outside the hostel in the dark having my morning smoke, waiting for Carlota and Jorge and doodling on my newfound walking stick. Looking down at my worn-out feet, I was delighted I could finally get my boots back on. With the wet day ahead and another predicted mountain climb, I was pushing into acceptance.

Together we began the first kilometer of 31. This would be the day we would arrive at The Cruz, the anticipation was churning inside me. The Cruz, translated "cross," was a meaningful place on the Camino to all pilgrims. All along we placed rocks in mounds also known as "*humillados,*" true symbols of the Way, and the Cruz de Ferro is the most significant. For me, it would be the place I would leave my past sorrows and express gratitude for what the Camino was giving me.

The gentle walk along the road at the beginning gradually turned into a narrow and uneven path. I was happy to have my boots on for having negotiated the wet rocky path in sandals would have been difficult. I was awestruck as the higher we climbed alongside the mountain the grander the views became. The slopes of all the surrounding mountains were in the full bloom of varying rich colors and the rising sun created the perfect light. It was one foot in front of the other in a meditative state. I had never known how serene walking could be for me. No music that day, only the gentle sounds of nature, the rain falling and my breath. We were all walking a fair distance apart in our own worlds, though each felt the connection to one another regardless. Finished with the first long climb of the day, we reached a rustic medieval village and followed the arrows along a stony road into it. Various remains of old structures, as well as newer ones that had been somewhat revived, flanked both sides of the road. I had read previously that this place was once almost deserted, being so desolate in the high mountains, until recently around 200 people returned to make their homes there. Finding a sign indicating food was available, we wandered through the big wooden doors. There were no other occupants inside other than a cheerful man welcoming us.

"Are we the last of the pilgrims?" Jorge asked the man.

"No, there are those few that come when the snow is on the ground," he replied pointing to the wall with photographs of pilgrims standing knee-deep in the carpet of white. Looking at the photos confirmed the man's words of the village covered in snow and it was then I realized how high we had actually climbed. We were seated in the cafe enjoying lunch when my ear picked up classical baroque, the Four Seasons played softly in the background. I had grown up with classical music thanks to my ballerina mother, who taught me the love for it. An emotional wave swept through me for the mother I had lost but had only now come to fully appreciate. Carlota spoke no words but reached over to hold my hand in solidarity with the overwhelming emotions that always flowed on the Camino. She knew regardless of how far we had traveled on our feet, it was the depths of our souls that had traveled the furthest.

Leaving the mountain village to resume the upward climb, the sun had come out to shine upon us and we gratefully removed our rain gear. The higher the Way led, the grander the vistas became that no single photograph could capture. As we passed a dense green forest on our right, Jorge said to Carlota and me, "I am to go in there...I will find you both later."

We both nodded in complete understanding as Jorge left the path and disappeared into the woods. I thought, as Carlota and I continued, Jorge was awakening to the spiritual mindset the Camino demanded. Feeling the new sensation, which had overcome him in the last two days, he had the desire to experience it further. The trees would comfort him and guide him to that very personal private time and he would emerge closer to understanding himself.

"He is seeing and feeling as we do," Carlota said, "this makes me happy not only for him but for us as a couple."

"Yes, I love the fact he felt the urge to go into the forest alone. He is such a beautiful person and watching him begin his acknowledgment of self is not something you often get to see in a person. Life does not allow for that intensity."

"Tess, you have shown us through your wisdom a different way to look at the world. We are both so happy you came into our lives, you are the ongoing gift from the Camino."

"Thank you, Carlota, and it is the two of you who are showing me what unconditional love can be. I came on this journey with a damaged heart, not only by Him but also by my own thoughts of what it should be. I am being shown every day what love really is and have restored my own hope."

Carlota and I had been walking in silence for some time, when we both felt the presence of Jorge emerging from the woods. I turned to smile at him and he returned the brightest smile before giving Carlota a loving kiss on the cheek. Whatever he went into the forest for was his memory to always carry in his heart and soul.

"Tess look!" Carlota pointed. "The Cruz."

I looked where she pointed and peeked out in the distance over the tops of the trees. I could see the iron cross reaching into the sky. The smoldering anticipation released because I would be there soon. Though still a few kilometers away, the path levelled out and snaked its way along the mountain, which allowed me to watch the landmark come closer with each step. The Cruz, from the movie I had watched eight months ago with Him, was my first connection to the Camino. It was portrayed in the film as a remote sacred place to release burdens. I fondly looked at my leather pouch hanging from my gypsy-made walking stick. The four stones I had carefully chosen and carried the entire way, from 7000 kilometers, away would be reaching an important destination.

Define the word expectation and it could simply state disappointment. I had lived most of my life in expectation: of myself, other people, places and, of course, love. Someone once told me, "learn to accept what is, to live in the moment and to remove expectations for then you cannot experience disappointment." Arriving at The Cruz, I immediately understood those words and the lesson that would follow. The anticipation momentarily dissolved into sadness and then disappointment as I surveyed the impossible before me. What I had expected was a remote place on top of the sacred mountain, only accessible to pilgrims walking the Camino. Instead I found the beautiful cross that had stood for centuries was nothing more than a tourist stop. We came out of the bush and into a large parking lot full of cars and buses unloading tourists that clambered up the sacred stones beneath the monument to have their picture taken. Further assault to my emotions were the municipal workers with bright orange garbage bags on the mound picking up the meaningful mementos left by previous pilgrims. I had but one choice to make at that moment and I clearly did it with ease. I would embrace that amazing symbol and what it meant to me and put all my focus into what I had come there to do, with that everything else no longer mattered. I bowed before the cross and the beautiful pile of stones from all over the world surrounding it. I felt a sense of release rising inside and then the desire to gently walk over the memories of so many before me to caress The Cruz.

Removing the stones individually from the leather pouch, in no order, I began my long-awaited personal ritual. The first stone, given to me by my spirit friend Kathy was to let go of the suffering in all our souls. The next stone was for Him, in hopes he could release the blockages of his life that debilitated him from being able to give and receive love and the third, a stone pendant He had given me long ago was laid to liberate me and to never settle for anything less than I deserved. The amethyst, the final stone, had been in my possession for a long time, I placed it down with a kiss goodbye and left all the thoughts of sorrow that did not belong in the future ahead. Taking a step back, I felt a nudge on my fingers and looking down it was the handle of Harold being slowly put in my hand by Jorge. Smiling, I laid my dear companion on the mound, the release of the crutch I no longer needed for the healing.

With the ritual of The Cruz behind us, having paid the homage needed, a definite gaiety and cheerfulness came in the air. Carlota knowing the importance of my experience asked, "So my friend, are you in your heart where you wish to be?"

"I am very much. I had carried those rocks far enough, much farther than I realized. My heart has changed to a new joy, no longer to be tangled in deceit and deception."

"You have used The Cruz very well. I too had the same new love inside when I placed my rock and today I stand with Jorge beside me."

"Carlota, I have opened my heart to love and who I choose to love first will always be myself. When I am ready to share this love It will be with the one that will never prevent me from being who I am."

KNIGHT OF TEMPLAR

While walking along the mountain top, I started to look behind me and marvel at the distance I had walked. I thought of my own West Coast mountain range back home that I had gazed at for years but had never thought to explore. There on the Camino, it was just a regular day as a pilgrim.

The trail narrowed again with one final small climb that I knew would be the last one of the day, thanks to my Camino GPS. Arriving at the crest, I saw there was an alpine meadow with a herd of caramel colored cows peacefully free-range grazing. After the crossing, the descent began gently as if the Camino knew to prepare our bodies for the downhill climb. I was acutely aware that the downs could be just as straining on the body as the ups and utilized my walking stick as best I could.

Carlota fell into step with me. "Tess, we will be coming soon to a very special place for me and I think for you too. It's an *albergue* like no other on the Way, I am excited for you and Jorge to experience it." I smiled at her in appreciation of the little things she knew about the Camino, having walked it before, thus ensuring Jorge and I wouldn't miss anything.

"Tell me," I asked her.

"It was built about 25 years ago by a man named Thomas. He is considered to be the last symbolically practicing descendant of the true Knights of Templar. The first journey of the Camino, made by King Alfonso II around 813, marked a growth of the pilgrimage and it grew so much that by the 12th and 13th centuries, there were thousands of pilgrims traveling the Camino every year. For that reason, other kings and clergy began to build *albergue* and hospitals to aid those on their way to Santiago de Compostela. This was not enough; it became necessary to offer the pilgrims protection as well for at times they were robbed of what little they had. That is when an order, famous in some ways but very hidden in mystery,

142

became the Knights of Templar."

I was intrigued by Carlota's story, appreciating the history and patiently waited to hear more.

"An interesting story, I will tell you about Thomas," Carlota continued. "When I passed by three years ago, I had the opportunity to sit down with him and he shared a story I know you will find interesting. As an ordained knight, there was a time he was feeling old and tired and no longer had the spirit to run the mountainside *albergue*, until the night he had a vivid vision. A woman had come to his doorstep looking for a place to sleep for the night. He told her he had no beds left and she had simply suggested she would sleep on the floor. He pointed out to her that the floor was dusty and cold, but this had not mattered to her. A place was found in the main entrance and she laid herself down to sleep. Concerned later in the night, Thomas went to check on the woman and she was not there. He walked outside into the moonlight, and, out in the field behind the *albergue*, he saw her dancing among the grasses. He approached her and she stopped dancing to withdraw from her red sash a mighty sword, which she then presented to him. He had been gifted with renewed strength and connection with his spirit. He took this as a direction to maintain his vigil for the protection of pilgrims and to continue his service as a Knight of Templar. Here Tess, are the real life stories of the Camino that are understood by only a few. I hope he is there and I can speak with him again."

Rounding the bend of an extremely narrow road, I saw the *albergue* of Thomas off the side of the road, built in levels against the hillside. The wind was blowing and catching the thousands of prayer flags strung through the trees around the primitively-built, stone structure. At the entrance was a handmade signpost with destinations around the world in kilometers. The one that caught my eye was Santiago 260 kilometers. Carlota and Jorge went in search of Thomas and I decided to hang out outside among the few other pilgrims that sat about. A very earthy man approached me with a smile deep and lingering.

"Hello," he reached out slowly to my neckline and lifted the tree of life pendant in his fingers. "Yes, I see this is who you are, you are strong and light."

I was slightly surprised by his forwardness but returned the smile, "Thank you, my Camino sister gifted it to me with similar words."

"You have a light about you that intrigues my spirit and I am very honored to meet you. Your eyes shine a blue I have never seen before. My name is Sven." I felt comfortable with this man, there was a uniqueness of serenity about him.

"I am Tess, thank you for your kind words."

"I stay here." He waved his arm, browned from the sun, around where we stood. "I help Thomas with the pilgrims' needs as they stop to rest."

"I appreciate you being here for us. It is a very peaceful place," I said.

"Will you stay the night here? I would like to share our histories together."

"Your invitation is lovely but there is more walking to do on my journey today."

Taking both of my hands in his and delicately kissing each one he replied, "Then go and share your light, Bella."

As Sven strolled off, I could not help but think that such a sincere kindness given from a man was a reflection of the newfound love I had for myself. I felt radiant and I am sure that is what he saw. Inhaling a deep breath of air into my lungs, absorbing how it is like the wind, feeling my body like a mountain and my mind, the sky above, I sat down on a wooden bench to just be. That was when Thomas drifted past. I knew without question it was him. He bore the presence of stature and history, wearing a white robe with the Camino cross on the front and at his side, a massive gentle wolf-dog followed in sync with his step. He disappeared into the *albergue* where I thought happily, Carlota and Jorge were waiting. I wanted to follow him but thought better to allow Carlota to have some time alone with Thomas. I sat peacefully, loving the sun on my face and feeling happy.

After some time, I entered the rustic, clay-walled *albergue* and saw Thomas talking with my friends. I approached the intimate group and the conversation paused as Carlota took my arm and eagerly introduced me to Thomas. His weathered face of many lives formed a serene acknowledgment of me, taking my hand in his for longer than a hasted moment. It was just another connection from the past to the present, feeling the power in his large callused hands. I stood in the presence of a being who was truly a symbol of spirit. Thomas gave each of us a Buen Camino blessing with a kiss to each side of our cheeks. Carlota, on behalf of our small family, spoke softly in Spanish of our gratitude to him for his protection. We left there all receiving individual gifts only for ourselves to know.

Not long after leaving Thomas I realized it was happening again. My boots had begun their deception once again with the balls of my feet burning in pain. With two more hours ahead to the next village, I quietly leaned heavier on my walking stick to soften the footfalls as best I could. The sun was shining in all its glory and I was sweating not only from the heat but the pain that was increasing. I struggled up yet another rocky path and slowly fell farther behind my companions. Reaching the top, they both stood there waiting with quiet concern, no one got coddled out there and from one pilgrim to another it would be insulting to do so. They just smiled at me and started walking again and, with my returned hobbling walk, I followed. On the next small summit, Jorge turned to me,

"What can we do to find you comfort Tess?" he asked delicately. This is how it had become between the three of us, just knowing when.

"I'm so hot and sweaty, I have too much on," was all I could come up with.

"Here let me help you to remove your pack and you can do what you need." Jorge offered with compassion.

"Thanks big brother, let's do this," I said with a weak smile."

Carlota offered reassurance that yes it had gotten very warm. Removing several layers of clothing, and stuffing them away into my purple monster, gave me some relief from the heat however the time we stopped walking only aggravated my stupid ankle as we started out again.

Carlota pulled out a solitary chocolate bar and split it amongst the three of us. "Sugar buzz to get us there," she laughed as did Jorge and I.

The smallest of comforts with love can encourage one to do anything. I plugged in my music for motivation to push the pain from my mind. The expression "mind over matter" rang true. At last, a short descent began towards the village that would be home for the night. But the path was over sharp, loose rocks and it was the final straw for me and what looked like for Carlota as well. My feet were on fire and my kneecaps felt like they would shatter into a million pieces. My pace slowed to that of a snail. Carlota fell into an accommodating step alongside me as Jorge bounded ahead, not for lack of caring he just knew we needed to finish on our own.

When we arrived at the mountain village Acebo Del San Miguel, population three in the winter months, the hostel was upon us immediately. Inside the comforting building, we were welcomed with enthusiasm by the host and without any formal paperwork handed keys to our rooms. Once sitting on my bed, pack dumped on the floor, I could not move fast enough to get the offending boots off my feet. Though the sensation of burning in my feet was slowing to a smolder, the freedom they now felt was wonderful.

"That's fuckin it; I am done with those stupid boots, no more," I said to myself. Slipping my bare feet into my beloved sandals, I popped back downstairs to bring a glass of *vino blanco* to my room. This had become my routine to unwind from the day and I no longer made excuses about it. I sat on my tiny balcony under a warm blanket, my wine, and my Marlboro's I and felt all good with my world. The view, for 10 euros, was breahtaking and in the setting sun I could see down the valley to the next leg of the journey. That was for tomorrow. I smiled.

SO LONG EGO

It had been a tough night, my right ankle was up to its old tricks again and had kept me awake most of it. The continuous hill climbing in damp weather from the previous day produced a swollen ,angry joint as I got out of bed that morning. Fortunately, the burning sensation in my feet was gone and I happily slid into my sandals. I gathered up my washed laundry, dried by the heater, and strapped the useless boots across the purple monster. Downstairs I found Carlota and Jorge having breakfast and by the look on their faces, their night was not much better than mine. We were all sore and tired and more tired.

Over breakfast, I brought up something I had been thinking about for a few days and wanted their opinion. When I started my research on the Camino I learned when I arrived in Santiago I would receive a Compostela, a certificate stating I had completed the pilgrimage. At the beginning of the journey, that certificate seemed very important to me, but, of course, things and thinking were ever-changing as I walked. I asked my family what it meant to them.

Carlota started, "Not as much as it does for most people. I didn't mind the first time I walked but now I walk with Jorge, for us, not the Compostela."

Jorge followed, "I do not care. I get it or I don't get it, it does not determine how I walk or when I choose to take "the other way," as we have talked about before. We make this entire journey for ourselves. You could walk the last hundred kilometers to Santiago, with proof of stamps in your Camino passport and receive the Compostela. So it does not have the same value."

"What! Wait a minute, you can just walk the last hundred kilometers and they hand you the certificate... I am so confused and don't understand. We have walked all the way from St. Jean—friggin—France," I said.

"I must also add," Carlota said, "the last 100 kilometers is not like the rest of the Camino you have experienced Tess, it is packed with tourists and locals, as well as the marketing souvenir shops that will give you a stamp in your passport if you buy something."

I was surprised to learn this information with no judgment intended. I could understand the thrill of those people accomplishing and receiving a Compostela. However, it really got my mind stirred as to the importance of that certificate to myself.

"Mmm, so I will have walked over 800 kilometers, crossed the Pyrenees mountains, fought many injuries, saw hell in my mind and thought through all the emotional difficulties but if I don't walk the last 100 kilometers I will not be acknowledged? Will that seems pretty awful. What if a pilgrim is injured five kilometers before the start of the last 100, all they have done will not count?"

"You are being silly little sister, it will count to you and that is all that matters," said Jorge.

"Yes, I see…" I paused, "many times in my life I have done amazing things with no recognition other than myself knowing it had been done. It has at times not been enough and I craved to be noticed and when I was not, it disappointed me. I know now I do not need accreditation from anyone or anything other than myself. I get it."

"That is a wonderful observation," Carlota said, "I feel the same."

"Our personal growth is our Compostela," I replied. "You know last spring I hiked into a steep foreboding canyon in Arizona. Which then I thought was very difficult and there was no one to share with other than my own feeling of accomplishment. I guess I was starting to learn the process then."

THE SIDEWALK OF HELL

Leaving the hostel to begin the day's walk, Carlota said to me, "Do you know you are helping us to start earlier in the morning?"

I laughed in response. "I am? Is that good or not so good?"

"It is very good, we are a little laid-back. One hostel we were at actually called us from the desk at 1pm and told us to get out. We laughed a lot about that."

"Well I have learned from you that 9am breakfast is okay. We have deliciously compromised," I said.

The most consistent recurring event on the Camino was, what goes up must come down. Directly out of the lodging we spent the night in, the slope down began through the village and beyond. My feet did not know what to believe anymore and it took some time to find my groove while accommodating my painful ankle. The path led around a huge horseshoe gully and I looked over my shoulder at the tiny village we had stayed at. It still amazed me how relatively little time it took to cross quite a distance. Jorge pointed ahead and in the far distance, at the bottom of the mountain, we could see our next destination 16 km away. Everything was within sight, the beginning and the end, it was very unique. The descent continued until eventually we went through a small village and agreed to not stop until the next village, a decision we would all regret three hours later. There had been nothing in between but a very long winding road built more for sports cars than for pilgrims.

Molinaseca, the town of the river, welcomed our already weary bodies as we crossed over the ancient stone bridge onto the main street. The first outdoor café we came to, I took off my pack and planted it in a chair.

"I am here," I said.

Jorge laughed and pointed a few doors down to an inside café. "And we go there."

The simple comfort of each of us wanting time alone without words or confusion was such an easy pleasure. After the unspoken rest time, Carlota and Jorge reemerged from the café as I was strapping the purple monster back on and once again we fell into step. The remainder of the day was a flat tedious seven kilometer sidewalk trek.

The sun was unseasonably hot and I was starting to melt under its rays. Having started the day high in the cool mountain, I had every possible layer of clothing on. I could see Carlota was feeling it as well. It was not possible to endure the heat on top of our already exhausted bodies and lack of sleep.

'Okay, we stop right here," I announced with a determined plan to drop my pack on the concrete. "I have got to take some of this shit off, you too Carlota, we are going to pass out." I threw my toque off, followed by my scarf, rain jacket, green Kermit fleece, and athletic jacket down to my long sleeve shirt, which I promptly pushed the sleeves up on my arms. Reaching down to my leggings I pulled them up and over my knees displaying my purple and lime green socks and sandals. This was not one of my more attractive moments, but it sure served us all a good laugh. Stuffing all my strewn articles of clothing back into my pack, and Carlota comfortable as well, we continued, still chuckling.

The light-hearted moment did not last long on the sidewalk from hell, as we were soon fed up with the hard surface, the cars, the impending noise and yearned to be anywhere on the Camino but there. Jorge tried to make jokes about taking a taxi but the laughs were not as boisterous as they had been when that joke had been previously made. We just slugged on silently into the dull approach of modern Ponferrada. Being surrounded by tall, sharp formidable buildings, people walking at alarming speeds, cars honking, and the smells that comes from a bustling city was a sharp contrast to the mountain we had just walked down. I felt—as we pushed our way along the busy sidewalk in our packs covered in dust, smelling of sweat—like we did not belong there. The looks from passersby did nothing to encourage our sore feet and beaten bodies. I caught my reflection in the store windows and was shocked how bedraggled and homeless I looked. Of all the populated places I had been, this was the first time I did not feel like an honored pilgrim, in fact very much the opposite. This was a lesson on tolerance, not only for myself but from the people in my surroundings. Our moods remained somber and I knew this had to change before we entered the old section of the city. Passing a large downtown café with inviting couches and tables I stopped.

"Let's stop for a bit, have a Coke and I… um, need a smoke, yes I really need to smoke," I said.

Carlota was immediately in cahoots with me and understood the idea of a reboot. Jorge was reluctant for the first time on one of our ideas, "Come on girls we are only 20 minutes away from the old city."

He did not look pleased with the idea of stopping; he just wanted to get there. Even with the awareness of how different our emotions flow on the journey, I still wanted to push the idea. "Please Jorge, just 10 minutes," I pleaded.

Jorge shrugged his shoulders and waved us to enter. It was a busy place with men in suits and women in high heels. My family sat down on a couch and I went to the counter to grab us some bottles of Coke. Returning to where our packs were piled up, I found Carlota with her feet out of her boots in Jorge's lap, getting a much needed foot rub. Lovely music was drifting down towards us from a young couple a few seats away with guitars. The atmosphere around us changed quickly as I handed out our cold drinks. No longer the loud sounds of the café or the traffic, just beyond invading our hearing, only the gentle sounds of guitars drifting in the air. Carlota closed her eyes and her lips were in the smallest of smiles as she relished in her foot rub. Jorge watched the guitar players with the sparkle back in his eyes. I was relaxed in the happiness of my two friends and well yes, I had a smoke. Twenty minutes later, our time in the café had lifted our spirits. Jorge was making jokes about Carlota's stinky feet and we were all laughing again. The reset button had been pressed.

"Coooome on, let's go," I jokingly whined. "We can't sit here all day, only twenty minutes to go."

I winked at Jorge and threw my monster on my back like it weighed three pounds. Carlota and Jorge laughed, and we headed out on the last part of the day's trek to what we loved the most, the old section of the city. Following the arrows around the last modern street corner into another world, my eyes widened, for there on a large mound was the first historical castle I had ever seen.

"Carlota, what is this?" I asked as we walked closer towards it. "It is amazing!"

"Castillo Templariode Ponferrada translated to Castle of the Knights

Templar. Ponferrada was first settled by Alfonzo Vlll around 1100 I think, in order to attack the neighboring area El Bierzo. Later, Alfonzo lX expanded the town and in the 1200s gave the town to the order of the Templar to help protect and defend The Way of St. James. The Templars took over the fortress and extended it for inhabitants and traveling pilgrims for protection. It was believed that the Castle of Ponferrada was the Knights headquarters for the region," she explained.

We continued into the depth of the old city in search of our lodgings for the night. Carlota had made arrangements on her magic phone to ensure we had places to stay. This had become a comfort for me that I was appreciating greatly. Carlota always arranged for me to have a private room, as I had shared with her the unpleasant story of my last night in the *albergue* dorm room. Knowing I was not a wealthy woman she would negotiate in rapid Spanish the best price. It also went without discussing that in the larger towns and cities we could stay in separate places, it was a sentiment we all understood and needed. Before dropping me off at my home, we all sat and had a glass of wine to toast our day.

"Jorge and I have discussed taking a day of rest tomorrow. Tess, you are welcome to carry on if you like and we can catch up later or enjoy yourself in this amazing place as well," Carlota said.

"Funny, I too have had the same thought. There seems much to see here, including the castle, and I want to not push myself to much."

"It has been very draining the last few days on all of us," Jorge added. "All for the good of course, but we all could use the rest and relaxation," he said, winking lovingly at Carlota.

"So we are all on the same page!" declared Carlota. We physically parted ways with a "see you when I see you" because that is how we rolled on the Camino. We all knew now that our connection was part of our journey and stronger than any plans, we would always find each other again.

I checked into the quiet room and was delighted beyond words to find a full-size bathtub, which I promptly filled and submerged myself in. It felt like years since I had one, remembering fondly the birdbath in Burgos that saved my hip. I was clean, warm and happy and regardless of the early time, I crawled between the crisp sheets and within minutes fell into a deep peaceful nap. A few hours later I felt refreshed and very hungry and, throwing on my only other outfit that was clean, I went out into the night to satisfy the growling in my stomach. I was not in the mood to sit in a restaurant so instead I went in search of a market. Eventually finding one, I loaded up on fruit, bread, sliced meats and a small bottle of wine.

As I headed back to my room, walking under an ancient stone archway that led into a narrow corridor with high walls on either side, the most evocative music filled the narrow space as I slowed my step coming upon the source. A man in a long coat with a brimmed Spanish leather hat stood alongside the wall playing on an acoustic guitar. His voice was haunting as he serenaded the Gaelic song of passionate sadness. I leaned against the opposite wall and slid down to sit on the cobblestone. The emotions of his playing and the way he sang stirred me like no other music I had heard before. The man sang his song, oblivious to those passing by who all missed the opportunity to really hear the fusion of guitar and voice. But I heard it and felt compassion for the sorrow he sang with the dark, expressive instrument. Never would I forget the mysterious man's voice and his guitar, it was the perfect end of my day.

LETTING GO

With the new day came the glorious thought of twenty-four hours without my pack on my back. One of the key lessons I had learned, and had begun to practice, was to listen to my body more, and though it spoke all the time I had rarely heard it in the past. So taking the day off from the Camino was an opportunity to rest, relax and enjoy my surroundings.

I spent the morning scoping out a dollar store and succeeded in finding rain pants, warmer tights and a turtleneck sweater as Carlota had mentioned the mountains ahead could be very cold and wet. I had also gone in search of covered footwear and with luck I found a solitary pair of camouflage-colored Reebok runners not only in my size but insanely on sale. As I cruised out of the shoe store wearing the new runners I thought of my hiking boots and decided it was time I would let them go. Other than being a heavy water bottle holder, they no longer served their purpose for me and I would find a place on the Camino to leave them. I walked past a fashion store and the new clothes smell once again drifted into my nostrils. It reminded me of the day I hobbled into Burgos feeling disheveled and in so much pain. With a "what the hell" attitude, I smiled and marched in to select a soft rose colored long sleeve shirt. I buried my nose in it and inhaled that much needed new shirt fragrance with the sales clerk looking slightly alarmed as I put it on the counter to pay. I headed out the door with my fancy bag containing a treasure. I couldn't wait to get back to my room to shower and put it on, so excited.

Later, sitting outside in a pretty café with a *vino blanco*, much earlier than I ever would pre-Camino, I felt fantastic in my new shirt and clean hair. Beside me were two plastic bags, one containing the laundry and the other with discarded clothes: the short sleeve orange shirt, carefully selected at home, the thin tights that rode across the Mesta with Nelson and, finally, the yoga capris that no longer kept my legs warm. I was conscientiously making decisions to carry only what I needed without attachment, something I had hopes of doing for the rest of my life.

After I finished my wine, I gathered up my bags and went in search of a laundromat, dropping my old clothes in the mission on the way. Strolling through the old city, I had noticed a change in appearances and sounds of the locals I passed. Having entered a new region of Spain over the last couple of days, I had observed the change in landscape and in the people. Their face structure was elongated, the skin color was lighter and their dialect sounded more crisp and sharp. I enjoyed looking at the beautiful people as I wound my way through the many little streets until I found a laundromat. Walking in the door, who should be folding her clothes but Carlota!

"Well hello sister. Of all the places I find us doing the most mundane of things," I said.

"Tess, it is good to see you." We embraced. "Your day has gone well?"

"Yup, check out my new shoes!" I did a little dance for her. "And a wonderful new shirt."

"You look good," Carlota said. "We spoke with Jorge's children this morning, it was so nice."

"Today is a great day," we laughed together. "Now show me how to use these machines. They're only in Spanish."

The two of us loaded my meager amount of clothing into the biggest washing machine I had ever seen. It was rather ridiculous but my only choice. Once the euros were inserted and the machine began its effortless job, Carlota left me to do my chores confirming we would meet outside the castle soon.

I sat outside on the step of the laundromat, embracing the sun on my face until the cycles finished. Once completed, I ceremoniously folded my tiny pile of clothes, bundled them up, and went in search of a pharmacy to settle a bet with Jorge. The wager was who had lost the most weight and for two euros I could stand on their scale. I was alarmed as, fully clothed, I weighed in at 115 pounds. I had lost five pounds since leaving home even while eating everything I wanted. A bonus, I also took the opportunity to get my height and strutted out the pharmacy all 5 feet and 5½ inches.

After dropping off my bundle of clothing, I set out to meet with my family to explore the ancient castle together. Remembering the lesson I had learned at the Cruz, I kept my expectations at bay and would allow myself to be surprised rather than disappointed. But there was no let down.

155

Walking through the magnificent castle, it gave me the unexpected throughout. The history inside expanded my curious mind as I took it all in. I strolled down stone corridors that had been walked by pilgrims centuries ago, climb steps that had been graced by the Knights of Templar, curled up in alcoves with small openings where soldiers had sat to protect the castle. The energy of history was powerful in that place of days long ago. I felt so grateful to have seen all that I had, as were Carlota and Jorge.

Before parting for the night, we savored a glass of wine discussing all we had seen in the castle. I brought up the Compostela again with them, as I was still wrestling with my thoughts on it. I wanted to share with them an encounter I had had with a woman that day.

"I met an older German woman today over breakfast who had walked from France as we did. She explained to me, with an exasperated sigh, that she had been to a doctor and he told her that her feet could not do much more and that she should shorten her Camino. Sadly, she explained that she was going to take a bus to Sarria, the last 100 km mark to Santiago so that she could receive the Compostela she so desperately wanted. There were tears in her eyes and she told me of her disappointment in herself, not to be able to walk the whole way. The best I could do was reassure her of what she had accomplished to date with her strength and power. I also reminded her that the Camino belongs to each one of us in our own way, and how it was done didn't really matter. She smiled through her tears and thanked me for listening. I felt sorrow for the pressure she put on herself to get the certificate."

"It is sad she will miss the beauty of the Galicia region in the mountains ahead," said Carlota.

"Yes," replied Jorge. "She will give that up to walk 100 kilometers with groups and marketing just so she can receive the credentials, it is very sad."

"I am losing my connection to that piece of paper," I said. "I trust the journey and myself to stay true to what I believe in. I will walk the beautiful region ahead, embrace the mighty mountains and stay connected to the Camino I know. When I reach Sarria, I would like to join you both on taking the 'other way.' It is more important to me to arrive in Santiago on All Saints Day with the two of you.

"And we want you to be with us too," said Carlota, smiling at me.

"So it is officially settled, we will finish our Camino together," said Jorge.

As I walked back to my room alone, I understood that by letting go of a plan and expectation, I would experience something greater than I could have imagined.

THE BOOTS

The following morning there was a sound I have not heard in 29 days, it was my phone ringing. It read no caller ID so I ignored it until it rang three more times. I was concerned it might be one of my kids so I answered but there was no one on the other end. I quickly message Sarah my daughter to ask if all was well at home, fortunately, she responded right away assuring me all was well. I felt reassured and just assumed it was a telemarketer as I packed up the purple monster.

I walked out of the hostel into the warmth of the brilliant morning sun and my family waiting. Together we left the beautiful city of Ponferrada, rested and happy to be walking again. Once past the city limits, we were again in the beautiful vineyards with leaves painted in every color of autumn. I remembered how when I started the Camino, the grapes were a heavy burden on the branches, begging to be picked, and now they were bare except Mother Nature's sign of the changing seasons. So much in nature had changed through the journey, just as I had.

The distance between us once again grew. I was unconsciously pulling ahead of my friends, it was a day to walk separately. It came from the comfort we felt between us and the unspoken need for our own individual time. Me, happily on my own, and them together, the connecting couple they had evolved into. Other than Jorge's walk in the woods alone, they had never left each other's side, it was a beautiful thing to observe.

Finished with walking through the vineyards, the path meandered into a forest of enchanting birch trees. I felt peaceful amongst the tall, white beauties of the grove. I came upon a young man sitting beside the path, next to a stream, selling a few pieces of handmade jewelry that he had laid out on a piece of leather. I stopped to look and was instantly drawn to a leather bracelet with a small, clay-carved Camino shell on it. I reached for it and he took it from my hand and tied it to my wrist with a smile only a man of the forest could offer. He accepted the random euro I placed in his hand, not a word was spoken between us.

Admiring my gift in the forest, I walked away noticing how truly wonderful my feet felt in the new runners for the first time. It was like landing on a soft cloud every time I took a step. As the forest parted to an open vineyard again, I glanced to my right and saw a three-foot birch stake embedded in the ground, slightly off the path. It was there I unceremoniously removed my boots from the purple monster and hung them, it was time. I walked away, leaving them hanging by their laces, hoping a fellow pilgrim passing by could use them. It was a fitting end to the attachment I had of them and freeing to no longer have to carry the weight. I'd carried them for a long time, it was the right place at the right time. Minutes later two large autumn-colored butterflies flew around me for much longer than one would expect and I had a strong sense of being close to my mom and my dad.

I arrived in a small town called Cacabelos a couple of hours later and I knew I was in a different part of Spain. The houses could only be described as having "old world charm" with clay walls, thatched roofs and pretty little gardens in front. The cobblestone streets narrowed the farther I went into the town toward a small, lively center. Finding a café in the tiny square across from the church, I drank my favorite Camino beverage, a café con leche, and drew on the delicious sensation of my cigarette. My thoughts had slowed down tremendously over the many days of walking and I was grateful I could just sit, admire my surroundings and think nothing or something. I was comprehending how little control I had over the past and the future, but by simply acknowledging the moment, I was thinking a thought from beginning to end.

"Where the hell are those two?" I said out loud laughing, as I ordered my third coffee.

"Lil sister we are here, you are there, and we are here… I mean, we found you… Or you found us…aaah so confused," Jorge said laughing when he arrived at my table with his arm dangling around Carlota's neck. "My woman… sit." Carlota who was also giggling sat in the chair held out for her and then Jorge plunked himself down beside her.

"Holy shit!" I laughed at them. "Are the two of you drunk?"

Carlota made the little "skosh" sign with her two fingers, "We are a little bit happy."

"That's so great, there has to be a story please tell me," I asked, clasping my hands together, "I am sure it is a good one."

"Well we practice the other way for an hour or so," Jorge muffled with his hand over his mouth so no one else would hear. To know this man as I did, his humor was undeniably funny at the best of times and under the influence so much more.

"Jorge let me tell the story, your English is not so good now," Carlota said as she placed a loving hand on him as he laughed from his belly. "First we must tell you we saw your boots and took a picture, that was so perfect you left them there. I am sure I can see you walking in the distance so you are in the picture, that would be cool, yes?"

"Maybe the cover for your book?" Jorge added, making us all laugh.

"Okay," Carlota continued, "So after passing your boots, on the long straight stretch that followed, there was a small white shed on the left side, did you see it, Tess?"

"I did, I remember two old men sitting outside as I walked by just before this town," I said

"Yes! That's it," Carlota smiled, she looked at Jorge and they are both now breaking out into a new laugh of the memory as I patiently waited to hear more.

"We were walking past the shed and Jorge asked the men how they are doing in Spanish and that was the beginning. They both stood up and invited us into the shed, which turned out to be a little winery and insisted we try several of their enology. As we sampled the delicious wines, one of the men showed us his pilgrim guestbook of the last eleven years of those he'd called in to visit. Tess, you would have loved it, full of dates and names from all over the world with personal comments about the Camino. I would have taken a picture for you but, I was on my third glass of wine."

I laughed at my beloved Camino sister enjoying the story, more so as she rarely drank more than one glass of wine, and appreciated the two of them truly living in the moment.

Carlota continued, "I was soon tugging on Jorge's arm and quietly saying that we must go but now he's thinking after all these days with us he deserves a little man time. I watched those three men converse in what I called a man's meeting. Reluctantly, after giving Jorge time, he said goodbye to his man friends and I think if I had not been there he still would be." It was then we all erupted into boisterous laughter knowing how true that could have happened.

"Now we eat," said Jorge. "I am going inside and seeing what they have."

"The two of you together makes my heart feel good and only Jorge would get you and him invited to drink free wine," I said.

"I know, I know," Carlota said shaking her head smiling," I am in love with him so much, Tess."

Jorge returned to the table with a tray full of different choices for us to share and we tucked in to satisfy our taste buds and hunger. As we ate, a large, kind-looking Spanish man came over to us and Jorge pushed his chair back to stand and pump the man's outstretched hand. Carlota turned to me smiling, "Ah shit, he is one of the men from the shed."
"Oh boy," I laughed.

The jovial man shared stories and, even though in Spanish, I listened with great content. The way he spoke, the movement of his lips and his facial expressions mesmerized me. It was obvious he was a great storyteller and Carlota was doing her best to translate the highlights to me. I am sure she wished I could understand the language better, I could tell there were parts that could not be translated.

We learned from him that all the church doors from here on would all be faced towards Santiago. Then they discussed the Cruz and how much it had changed over the years, though still symbolic, its modernization was obvious. I was relieved to hear this thought from a Spaniard, who, I understood, had walked the Camino many times. The following story he told us was a shocker, especially for me as I had a distant plan to make it part of my Camino. I had discovered, through other pilgrims, that a town 60 km past Santiago, called Finisterre, referred to as "the end of the world," rested on the cliffs overlooking the Atlantic Ocean. It was there many pilgrims would officially finish their Caminos with a ceremony of burning their pilgrim cloths to symbolize the start of a new life. That kind of thing intrigued me, but sadly the man said for me not to bother for, like the Cruz, it was overrun by "burning" pilgrims. It had become a trampled tourist town that relied on the burning tradition and ocean to draw people to it. He further told us that the accurate end of the world is the coastal town of Muxia, which is the actual western tip of Spain. It could be reached with an additional two days of walking from Finisterre and was "the place of wonderful energy, striking beauty and a connection to freedom."

That was the start of me rethinking my plans as my eyes settled on my tattoo, "Trust the Journey." He continued speaking, and Carlota gave me the translation, as she knew I was intrigued to know what he was sharing.

"Centuries ago in Spanish history, Finisterre was known as the 'land of the dead.' It had always been a very poor area on the coastline and fishing was impossible for the sea was unfailingly rough and angry. The villagers would set up a light source, at the same point the pilgrims now burn their clothing, at night along the jagged rocks. It would encourage ships, loaded with men and cargo, to sail to their deaths into the large sharp rocks. The villagers would then raid what cargo they could retrieve and scurry back to the village," the man explained. I couldn't help but be amazed at what I was learning and I was grateful for the man that he had shared this strange history. Upon completion, the man stood and wished us all Buen Camino and disappeared around the corner.

"Wow, that was so interesting. Thank you, Carlota, for translating for me. I would have missed such an incredible story," I said.

"It is no problem, I know you enough to understand your desire for these kinds of stories," she replied.

"One of the great moments that happen on the Camino, plus free wine," Jorge laughed.

Finished with our lunch, and all feeling the need to get going, Carlota informed me that they were going to go to the pharmacy to get some supplies.

"Okay," I replied, "I'm going to go ahead and start walking." I felt like continuing my day of solitude.

"Let me just tell you," Carlota said," Outside of town you will come to a fork in the path with no markings. For some reason, there have never been arrows placed there. Just stay to the left, if you go to the right it is another hour of walking."

"Thanks for the heads up, see you later."

DEATH

It was a long climb out of Cacabelos that led to the fork in the path. I went left, as instructed by Carlota, but an unexplainable pull brought me backtracking and taking the path to the right. My thoughts were if I have walked this much and this far, no reason to question why I wouldn't want to go the other way.

I was very much on my own walking through several kilometers of the grandest wine vineyard of all. If I had thought the colors of autumn could not get more vibrant, they did. It was like I was walking through a Monet masterpiece with movement in the soft wind. The vines seemed to part to allow me to pass through the expanse of the orchard. Coming to the end of the comforting section, the path started a slow descent towards a small village. The moment I crossed the invisible boundary, an unwavering feeling of death overtook me, I was suddenly very cold. Entering the village, it looked as if it had experienced a disease, resulting in the appearance of the demise. The smell in the air was nothing I could compare to and I was surrounded by stillness, nothing moved. The decaying homes were built from stones and would entice me to wonder why there was no life there. For the first time on the Camino, I felt uneasiness and out of place. My inquisitive mind could not help but peer through the dust-covered windows of some of the houses. From what I saw was, aside from dirt and cobwebs, the occupants had left suddenly. It was not fear I felt, only a strong sense of being very alone, I had walked into another world.

Leaving the village behind me, the Way opened up into the warmth of the sun again. Out of the darkness, into the light. Briefly, the question in my mind was why I needed to experience the village, was that why I went right instead of left? With no answers coming, I shook off the desolate feeling, plugged in some music and sang for all of nature to hear.

Hours later, I arrived on the outskirts of Villa Franca, a town nestled halfway up a hillside between two mountains. I remembered reading some history of that place. It dated back to 776 and was considered the beginning gateway to Santiago. The excitement and yet sadness started formulating in my chest as to how close I was to my destination. The third stage of the Camino had begun. Coming in closer, I could see Villa Franca was built in layers on the hillside and from where I stood, the town zigzagged its way to the bottom. As I entered, I came across a 12th-century church with the well documented La Puerta del Perdon, the "Door of Forgiveness." Dedicated to St. James, it is the only other holy place, other than the Santiago Cathedral, where pilgrims could receive absolution for their sins. Though the church was not open, I stood beneath the arched doorway to embrace the symbolism of the blessing. I was following in the same steps of the centuries of pilgrims before.

I carried on into town to find the house of sevens, the hostel Carlota had arranged for me to stay for the night. She told me I would find it easily, as it sat across from a staggering fortress, and she was right. Los Marqueses del Bierzo stood impressively in all its massive stonework above Villa Franca. What defined this awestruck fortress was its architecture of tall stone walls with gunnery slots running all along the top and the massive turrets on each corner. The large impressive double doors of the entrance, made from a dark wood, appeared to be the only way into the fortress. Carlota had told me there was not much public history and, unfortunately, it was privately owned, so it was not possible to see inside. It did not matter, as I could savor its history and magnificence from the street in front of my home for the night.

I dropped off my pack in room number six, out of seven, and set off to explore the town. I meandered my way down the hill, enjoying the painted front doors in every imaginable color, flower boxes on all the windows and the people all smiling and happy. My natural homing device led me into the main square full of activity in the late afternoon sun. After a short stroll through the square, I settled at a table and was quickly served a coffee. Beside me sat a very striking looking man who acknowledged me with a smile.

"Camino?" He asked.
"Yes," I replied.
"Where did you start?"
"St. Jean-Pied."
"That is good, that makes for good Camino. I too started there," he said.

I felt brave and wanted to know more of this man and for the first time I asked the question, "Why are you walking?"

"Just wanted to do it," he paused for a moment, "not everyone has a tragedy or needs something from the Camino. However, there are things I have learned that I will always keep with me."

"I have also learned that many things will always be with me."

He gazed down at my sandals and chuckled, "You walk in those?"

"I have been for a few weeks, it is a long story, but the end result my boots are hanging off a stake in the ground along the way."

He smiled at my words and replied, "I also abandoned mine," he pointed to the boots on his feet, "but they found their way back to me, go figure."

"Maybe I will see mine walk by me one day!"

The two of us had a laugh about our boots when Carlota and Jorge pulled up chairs to join me. I took a last look at the interesting man and wished him a good Camino, he nodded his head and I turned my attention to my family. We exchanged stories and experiences of the day and how much we love the Camino world. We discussed how the third stage had begun in our mind and spirit and how we could not imagine not being on this journey anymore after it was finished.

"There is a marker outside of my hostel, it says 180 km to Santiago. I cannot believe we are this close." My eyes widened at the thought.

"Yes, I know it's incredible and the walk ahead is the true Galicia region. The scenery will be breathtaking most of the way," said Carlota.

Our drinks arrived and, as Jorge was paying, a tall, one-toothed, long-haired Portuguese man came over to our table. He spoke in Spanish to Carlota and said he could see a bright light of energy coming from me. Carlota was translating for me as he spoke.

"I could see her from across the square, and now the two of you have arrived and the energy is so large I had to come over," he said. Carlota was intrigued as was I and he continued on in Spanish while Jorge rolled his eyes.

"This is who Carlota and I are, we care for everyone and want them to feel happy. If we were in Canada I would be doing the same thing." I giggled knowing Jorge well enough to know he needed an explanation. The strange, tall man looked at me and indicated he would like a hug, seeing no harm I stood to hug him. When he went to kiss my hand Jorge put a stop to that with the intention of protecting my honor that he would explain later. The man smiled and went on his way.

"It is my job to keep an eye on you." Jorge winked at me. Carlota smiled in agreement for she too saw the man in the same way. "I guess it can come in all forms... Not just trees?" Jorge shrugged his shoulders.

"You bet big brother, eyes wide open," I said.

Another Camino experience we all agreed on, before saying good night to each other, with plans to meet in the morning. I was hungry and funds were definitely running tight so I went in search of a market and quickly found what I needed. Returning to my hostel, I wandered into its tiny restaurant to ask about a corkscrew for my bottle of wine.

Annette, the woman behind the small bar area was ever so kind to me, opening up my wine and asked, "One glass... Or two?"

I had gotten used to that kind of question, being solo, and now I just smiled and replied to Annette, "One please."

"For me, that is a wonderful thing. A woman who can enjoy a bottle of wine alone has her thoughts sorted out and is in a good place within her own heart," Annette said.

"Thank you and, you know, I think I really do," I replied to her memorable words. In my room, I devoured my array of snacks, relished in two glasses of wine, had a shower, wash my clothes, hung them to dry and snuggled into bed.

THANK YOU

The following morning, I sat in the hostel's little restaurant, gazing out the window at the fortress, feeling very lucky to be where I was. I was giving some thought to Him, though no longer feeling the attachment I had when I started the journey. We had messaged on occasion, and he had expressed a genuine concern for my well-being, as well as his regret for his past actions. I had long since forgiven him and told him so, but the trust would never recover from how much he had hurt me. I could only hope he would find a way past the sorrows he carried and live a life where he could feel love.

It was a quiet walk through the ground-level forest path. The three of us walked closer together, Carlota and Jorge hand-in-hand just ahead of me. I was accustomed to observing the two of them, like a slow movie reel revealing a story. I was comforted to see the love and it reminded me it is possible with the right connection. A few days prior, I had shared with Carlota and Jorge one of my favorite songs of love: John Forgetty's Joy of my Life. Not only did the lyrics suit them both perfectly, but the tune itself rang out their personalities. They had loved it and would keep that song as part of their memories of the Camino.

The afternoon had brought us to the village of Las Herrerias, a tiny community at the base of the largest Galicia Mountain. I looked up at what I would be tackling the following day and embraced the determination I felt. Jorge and Carlota had left to go rest in their room and I settled myself on a stone wall that meandered the length of the narrow roadway just outside the hostel. The wall enclosed a herd of peaceful black-eyed cows grazing in the setting sun. I just watched them. Their sense of tranquility and quietness encouraged me to do the same. The moment could have been anywhere in the world in any given century and I was comfortably lost in it. The sun shining on the yellow and green Galicia mountains in front of me was an invitation to come to climb and experience its magic. I could hear songbirds singing and happiness and faint Celtic music playing from somewhere unknown in the distance. My mind reflected on the past few days and all I had experienced: the gypsy village, the Cruz,, the hamlet of death, the Portuguese man, and so much more. They had all connected me to my spirit that was waking up in full force.

I was in the last stage of the three transitions of the Camino, spirit. In three or four days, I would be walking into Santiago and then what? This had been my life, I could not possibly imagine doing anything else. What I had was of two emotions; elation of completing the incredible pilgrimage and the sadness of leaving. Acknowledging these feelings, I refocused my attention on the gentle, peaceful movements of the grazing cows to bring me back to the present I was in and just be.

Sometime later, I went to my room, thinking of all the rooms I had called home. Finding that feeling of comfort in every place I had stayed, for it was my own presence that made it home. Tackling my evening routine of showering and washing clothes, I no longer did it out of thought but out of habit.

Donning fresh clothes, I ventured out to the hostel's outdoor tables with a glass of *vino* in my hand. There was something I had to do and it was time. I called his number in Canada. With the time difference, I knew he would be asleep so I was prepared to leave a message from the Camino. Hearing his voice on the recording stirred my heart and memory of Him and with a cleansing breath, I began:

"Hi, it's me. I started this pilgrimage because of your challenge so many months ago and I want to say thank you. I have learned more about the woman I am in the last 28 days than I have in my entire lifetime. I ask you to try and do the same for yourself. I saw a light in you no one else did, but you blocked it with sorrows and anguish that constantly interfered with your own inner happiness. It was unfair of me to think I was capable of helping you by just trying to love you. I know now it is something you must want to find and you must do that on your own. How you do it is entirely up to you but please do it. You will always have a piece of my heart, which I leave with you willingly. Draw on what I gave you because you are in your own way a good man. Simply open yourself to the journey of life with eyes wide open and I promise it will bring you what you need to find your piece. I wish this for you, take care."

Pressing the end call button, I knew then why he was part of my own journey. I felt I had completed something, having done all I could do to show him the goodness in the world, and it was then I let the negative energy of us go. Had I not experienced the time with Him, I would never have been forced to look within myself as I was now doing.

Listening to the Celtic music playing gently, soothing my soul, I thought of the best gift the Camino had brought me, the very best lesson, love. The definition of love between two people was visual to me everywhere I went but most predominantly in Carlota and Jorge. They walked together in step, laughing, talking, crying, all in beauty. They privately, without words, looked after each other, in a way only I could see. The tender Spanish conversations of endearment they shared resonated their love, a universal language that needed no translation. The way they no longer left one another without a touch, a look or a soft kiss. The sparkle I saw in their eyes when the other was telling a story, and the acceptance and understanding with humor of their individual shortcomings. The instant recognition when one was suffering physically or mentally, and the ability to guide one another into healing. I was witnessing it all and would continue to do so all the way to Santiago. The two of them, without any effort, gave me what I'd been seeking my entire life, to see and to know what love really was. To hear it breathe, to watch it grow, to see it flourish and admire the unconditional expectation of it all.

I looked up then to the window and saw Jorge at the inside counter, I went inside.

"All good big brother?" I asked.

"Aw my girl, she is not feeling well. I have come to get her something to eat."

I smiled at him remembering the thoughts I had just had, "You are a good man Jorge, please give her my love."

"I will always do for her whatever I can, always," he replied, gathering up the prepared food and disappearing up the stairs.

I also needed to eat and sat down in the small dining room and ordered a good ole hamburger. My main diet had been fruits, nuts, and tapas for most of the Camino and suddenly I was craving beef, must've been the cows I chuckled to myself. I was served the biggest hamburger I'd ever eaten, and had to do so with a knife and fork. Devouring the entire thing was not a problem, knowing I had a mountain climb ahead, I thought it was a good investment for energy. Following the burger came the most decadent piece of chocolate cake. It was passion at first bite and I washed down each mouthful with wine. Heading to my room with an additional slice of the divine cake, and two forks, I knocked on their door. Jorge answered smiling.

I thrust the plate into his hands, "You just gotta trust me," I said and scurried down the hall.

FINDING UNICORNS

Gazing out my sun-filled window, I was happy the predicted rain yet to arrived. Curious, I pulled out my phone to have a closer look at the forecast. Temperatures would reach 24 Celsius with no rain all the way to Santiago. This was not the weather that had been forecasted several days before. In Ponferrada, I had loaded up on warm clothes but, funny enough, there was a possibility I wouldn't need them at all. Carlota had suggested numerous times I would need a poncho for the severe rains but I couldn't seem to get my hands on one. Well, apparently I was going to finish the Camino without needing one, ta dummm!

The three of us reunited that morning over breakfast.
"Hey sister, are you feeling better today?" I asked Carlota.
"My God yes, fantastic today! I am sure it was the chocolate cake you brought us."
"There was no way I could not share the most amazing taste explosion with you," I laughed in response.
"We really enjoyed it a lot." Carlota smiled at Jorge who added, "Yep, that was a good idea."

Finishing our breakfast, chatting about our anticipation, we heaved on our packs and set out for the day's climb. The sun was already hot and having started out the "other way," it was already after 10 AM. The incline began on a narrow asphalt road that reflected the heat, making the already laborious climb tougher. Eventually, the road disappeared and we climbed a dirt path amongst the trees. We walked through an umber brown forest that reeked with age. The woody incense came from centuries of snapping branches falling to the forest floor to rot slowly. Every sprawling tree we passed under reminded me of a watchful guardian. We ventured deeper into the tangled heart of the primeval forest expecting secrets to be revealed around every corner. The farther in we walked, the more mystical and spell bounding it became. Huge roots spread eagled along the ground, twisting and turning in all directions.

"There are chestnut trees," I broke the silence, "I see chestnuts everywhere, and listen." We paused to hear the sound of popping as they simultaneously fell from the trees. Picking one up I admired the thorny shell, which had encased the treasure before it hit the ground.

"To me this represents the mightiness of life," I said, holding up the chestnut. "If the world is destroyed, this would survive and the seed of growth would start again."
"If that happened," Jorge said, "we will meet in this forest."

I smiled because I knew he understood. The foliage became thick, forming a fairytale canopy above our heads. Arthritic bows, gnarled with age, dropped their bounty of not only chestnuts but also walnuts. At any moment, I imagined I would see a fairy peeking out from behind a rock or a unicorn behind one of the great trees.

Leaving the magical forest, still on the upward climb, the path brought us into a quaint hamlet. Walking the only dirt path entrance into the center, we came upon the modest café. Dropping all three packs on a bench, we ventured in for a beverage and, to Carlota's and my delight, there was a display of metaphysical trinkets and jewelry. While Jorge took care of getting us drinks, Carlota and I looked over all the items. I was drawn to a collection of Buddha prayer beads made from seed shells, each shell was carefully threaded with twine. Three tassels hung down the same color as the shelves. There were two that looked exactly the same but each held their own energy so I discreetly paid for both of them. Outside sitting on the bench, sipping our drinks, I took them out of my pocket.

"One for you and one for me." I draped one around my Camino sister's neck and one around my own. "A day to remember."
Carlota smiled in understanding. "Oh Tess, I love them, I do and feel they belong to us, thank you."

I grabbed my pack, hugged both my family and silently headed off down the path, we could easily do that when we needed it. I wanted to walk alone and listen to some Enya, music that belonged there in the mystic mountain. Leaving the hamlet, the path continued up. The forest widened and the valleys far below became visible. I could see a great distance and I was amazed how high I was. I appreciated my powerful legs for having brought me here, although the tedious footfalls had intensified the pain in my old, injured ankle again. No shoe was going to prevent the pain that encased my joint. Sudden disappointment overwhelmed me and I grimaced in my steps while leaning heavily on my walking stick. Why must I suffer more? I did not understand.

In exasperation, I looked up to seek comfort from above but instead, standing on a small mound, silhouetted against the blue sky, a unicorn stood. He lifted his head from grazing and looked at me with his large black eyes. Without a thought, I scrambled over the small stone wall covered in thorny bushes, I needed him. Running my hands over his plump sleek coat, burying my face in his neck to inhale the aroma I knew so well, horse. He in turn, swung his head over my shoulder to return the love I was giving him. I had a solitary candy in my pocket and I offered the tiny gift to him to say thank you. His velvet lips graced my hand as he graciously accepted. New strength and power coursed through me, I knew I could get past any discomfort. With a final hug to the magnificent creature, I returned to the path and realized every time on the Camino that I was in a difficult moment, needing a strength I could not find, horses appeared. The Pyrenees, the industrial area outside Zubiri, the gypsy village, and now here in the Galicia mountains, they were my earth angels. I glanced over my shoulder for a final look at my friend but he was gone.

"Thank you," I whispered.

The final ascent was extremely steep and uneven, but with views that were unimaginable. The landscape had changed so much over the course of the journey. I'd experienced so much variation in the beauty of Spain. There in the mountains, the connection to nature was more than I could've ever conceived. As I stood on a ledge overlooking the view, I was in awe at the beauty of this amazing world. The emotion of spirit, the Universe, and the God of my own belief once again overcame my awareness of how nothing less created the greatest miracle of nature. I spent time there alone in the amazement of my surroundings and breathed the air deep into my lungs, quietly thanking myself for being there.

Not long after the sacred place, I came across the official marker indicating I was now in the Galicia region. I had one amethyst left, one that I had carried in my pocket for years, long before the Camino. Taking it out, I laid it on the stone marker. It felt right to leave the last part of my old self in that place. The Camino gifts and my gifts were completely intertwined now, and it was there Carlota and Jorge rejoined me and together we completed the last haul to the top.

The village of O Cebreiro, burrowed in the mountain peaks, greeted us. The sense of traveling back into a Gaelic time was felt by all three of us. The church, facing toward Santiago, was the first to entice us into its walls. It was built out of far smaller granite blocks than I had seen before, and, once inside, the simplicity provided a place to pray. There were only a dozen pews in the small church and a large crucifix adorned a wall but that was it for decor. I proceeded to the side of the church to perform a Catholic ritual I had not done since I was a child. I took a birch stick and lit it from the center candle, a little votive for my mom. I had made peace with her on the journey and I missed her. Without a doubt, I knew my mom was proud of me connecting not only with myself but also with my European background that she held so dearly. Knowing she felt this way, gave the peace we both needed. Carlota took a card from a small table with the Gaelic pilgrims prayer and handed it to me, I would read it in silence later as the words were gratefully printed in English.

O Cebreiro presented itself as another world in another time. The first thing I noticed was the hue of everything from the houses to the cobblestone in the street; everything built using small river stones was every imaginable shade of gray. Each house and building, all with thatched roofs, were discolored from the sun into the same variations of grace, the only splash of individuality to each structure was the front doors painted in vibrant colors of greens, reds and blues. It was a charming place and I enjoyed looking at all that was made from what nature could offer.

"Guy's we have a problem," Carlota announced after ending her call. "There is no room here for us to sleep."

"That's too bad I would've liked to stay here, so much to explore," I said.

"Yes us too, but the bigger problem is we still have quite a walk to the next place, and I should start making calls now," suggested Carlota.

"Well girls, let's stop and have a late lunch," Jorge said. looking at Carlota, "And you can make the calls."

Finding a small café with indoor and outdoor seating, Jorge and Carlota went inside and I chose to sit outside on stumps pulled up to stone tables, I liked that. When lunch was finished Carlota and Jorge came outside as I was finishing my smoke.

"So Tess, I have found a place, but we must walk another 12 km!" Carlota said. We had already been walking for seven hours and we all knew without a word that we would have to get going to beat the darkness.

"All right then, let's go. I'll just pop in and pay for my lunch," I said, heading in the café.

The woman behind the counter smiled at me as I approached. "You're very nice companions have paid for your lunch," she said.

"They did? That was super nice of them."

"Did you enjoy the lunch I made for you?"

"I did very much thanks," I replied.

"Very good. I wish you a good Camino."

Once I was back outside with my family, I thanked them for lunch and what a treat it was. The two of them were muffling laughs and I had no idea why.

"What's so funny?" I asked. Jorge took the liberty to explain, laughing, what had happened between me and the woman inside.

"We could hear you talking together and the whole conversation she spoke to you in Gaelic and you were responding in Spanish!"

"What? Come on you are kidding me," I said.

"It is true," piped in Carlota. I laughed too at what had just happened, I had no idea I had just done that.

"Well, then I must be turning into a Spaniard."

We all enjoyed the laughter as we started the next 12 kilometers.

Not long after leaving the village, I felt a creeping regret in my gut about what I had chosen to eat for lunch. I'd forgotten my personal commitment to never eat fish unless near water. Shit… My stomach was rearing its ugly side as I tried rubbing it to settle it. To add to the discomfort, the Way was taking us up another steep, steady climb. Silently the three of us hunkered down, heads bowed to the hill, packs rolling over our shoulders, walking sticks digging into the ground, working hard for the climb. Jorge was several feet ahead, Carlota several feet behind, when I stopped for a minute and shouted out, "REALLY?"

That word had become a sort of mantra for when the going got tough, and it brought all three of us into laughter on the inclined path.

Jorge called down to me, "I was thinking in my mind any minute… Any minute she is going to say it." More laughter erupted, easing the focus and resolving the stomachaches for me.

Once again we turned our attention to the climb and, as with all climbs on the Camino, the discomfort endured was greatly rewarded. Upon reaching the summit, there, silhouetted against the setting sun, stood the largest and most compelling metal statue of a pilgrim facing towards Santiago. Over 40 feet in height, it was breathtaking as the sun rays filtered through the gaps of the outline. These are the markers on the Camino that remind pilgrims how incredibly strong they are. We each took a turn having a memorable photo taken with our large fellow pilgrim.

As the sun disappeared, we walked away from the statue as still more kilometers before darkness would rob us of our vision. I was feeling the chill of the dusk air in my sweat-drenched clothing as we set the pace as quickly as our legs would allow.

We arrived in a settlement, after descending from the statue, with four kilometes still to go. Carlota suggested she could phone the albergue and ask if someone could pick us up. It had been a long walk and we all felt the day weighing on our bodies and with the darkness, it was agreed. Carlota made the call and informed us that as soon as the *albergue* owner's husband finished bringing in the cows, he would come for us. We had to giggle as only on the Camino would we be waiting for cows to come home. Carlota was given instructions for us to wait outside the one and only tavern in the tiny community. We ventured up the narrow dirt road in search of it, dodging cow pies and passing other cows all tucked in the barns for the night. Finding the agreed meeting point, we sat down outside with our backs against the wall and waited.

I was gazing down the road when I noticed two other pilgrims walking towards us. As my eyes began to focus, I jumped up with excitement.

"Eva!" I shouted, which grabbed everyone's attention.

"You are kidding me, Tess!" The two of us embraced.

"Of all the places we find each other. Here." I swept my arm around the almost nonexistent village.

"This is crazy, yes?" Eva replied. It was then that I noticed a light in my friend's eye, so different from the last time I had seen her. The man standing quietly behind her was sure to be the reason why.

"Tess, this is my husband, Alex." She stood aside to allow him and me to shake hands which quickly turned into a hug.

"It is such a pleasure to meet you. Tess, I have heard so much about you from Eva. When you jumped up, I knew it had to be you," he chuckled.

"Oh Alex, it is so wonderful to meet you as well. You, joining Eva is the best thing for my eyes to see."

"And of all places, we meet behind a hole in the wall tavern, crazy Camino stuff," Eva said.

I introduced her and Alex to my Camino family. In the brief conversation, Alex became everything Eva had talked about since our meeting in Zubiri so long ago. I loved that he was now with her to walk together on the final days to Santiago. This was a different kind of love story than Jorge and Carlota's one of new love, it was one of long love.

With a final farewell, I turned to hug Alex and said, "You are a man I can honor with what you have done for Eva."

He quietly responded to just me, "If you hold something too tight it will die, give it the air it needs and from that it will become something greater." He looked over to his wife with so much love, took her hand, and disappeared into the night.

"She is a super nice lady Tess, I understand your connection. From the stories you have told us, she is the person you have known the longest on the Camino," Carlota said.

"Yes, it is amazing how many encounters we have had, all brief, but we have learned so much about each other. The value of a moment has come to mean so much to me."

Resuming our waiting against the wall, Jorge said to Carlota, "It has been a long time waiting for the cows, yes? They must be very slow tonight," he laughed.

"Yes it has," we both replied.

"I will go inside and see if I can discover anything, and then maybe another phone call," said Carlota, as she disappeared into the tavern. Coming back out moments later, she said, laughing, "There is a door on the other side. Our ride has been having a beer with his farmer buddies while waiting for us."

"Oh my god, you are kidding? Really?" Jorge smiled, slapping his leg.

We grabbed our packs and cut through the bar to find the man waiting by his car,. He told Jorge, man to man, he appreciated us being late so he could have a beer with his friends.

I turned to Carlota once we were settled in the back seat, "It still amazes me how things work out on the Camino; if we had not made that mistake of waiting outside that back door, I would not have seen Eva and Alex.

"The Camino provides," was all she said in response.

"So very true."

Pulling up to the private home after the brief car ride, we were immediately sat down and served cold meat from the farmer's cows, various cheeses made from the milk of his dairy cows, and homemade warm crusty bread. With a couple glasses of wine, the three of us felt like royalty and a little tipsy, laughing at everything about the day, especially the unknown second door of the tavern. Exhaustion finally overtook us and we bid goodnight with thank yous to the man and his wife for the generous hospitality, and went off to our rooms.

As I lay in bed, I thought about the incredible day behind me: the mystical forest, the mountain climb, the "unicorn," and the victory of holding my walking stick in the air on the peak. It had been a good day. Drifting off to sleep, I was thinking of the many aspects of my final days on The Way of Saint James and what my life would be like after the Camino.

THE BEGINNING OF THE END

I woke with a surprised feeling of grief in my heart. The end of my daily routine on the Camino was drawing near and I was struggling to accept it. In life, we dislike losing something we have come to love with all that we are, but it is part of the journey of always moving forward.

Reluctantly, I got out of bed, packed up my purple monster, and headed down to the smell of breakfast coming from the small dining room. I sat alone, aside from a quiet couple at a table in the corner. Observing them discreetly, they pulled me out of my slump. They sat so close together, it appeared that they were of one body and, without words, communicated with smiles and touch. They were a contrasting couple. He looked younger with beautiful Portuguese features and she had rosy cheeks against alabaster skin. The reason for their sacred silence became clear to me then, he was deaf. Their small hand movements spoke of the love and adoration for each other, it was a beautiful sight.

Checking if wifi was available, I pulled out my phone, sent a message to Sarah to let her know I was doing well, knowing she would share the message with her brothers. I missed my children. They had unquestionably stood behind their wanderlustful mother every step of the way. There too was James, who had been my friend through not only the beautiful moments but also through the fears and trepidation. I decided to facetime him to again confirm my thoughts on my plan for how I would finish the Camino.

He picked up quickly and I could tell he was delighted to see and hear from me.

"Hi you," he said first, "What a nice surprise."

"Hi to you too. Sorry for calling so late your time, but I'm sorting something out."

"You bet, I am here for you Tess."

"I am still challenging my thoughts on skipping the last leg into Santiago. I feel a pang of guilt that I am not doing the Camino to its fullest."

"I hear what you are saying, but had you not already told me that you are doing this to avoid an area that will not be what you have come to learn is your Camino?" he asked.

"Yes, that is true, but the guilt I am feeling really sucks."

"Guilt to who Tess?" he challenged me. I had to pause and think about that question.

"I think it is the guilt I am putting on myself about people at home who think I am jamming out or taking the easy way."

"First off, that is ridiculous to think after how far you have walked and what you have experienced, that you are, as you say 'jamming out.' Second, my friend, it is none of your business how or what other people think of your journey," James said, wagging his finger at me on the screen. "It's not up to anyone but you to understand what you need to do. The people that love you, really love you, will never question you."

I responded smiling. "Right, it is a good reminder. In fact, if I take all the lessons I have been learning this is where it all comes together. It is my life to live and I am under no obligation to explain myself to anyone but myself."

"You got that right girl!" He said.

"And if I stay back and walk those other kilometers, I will lose the opportunity to embrace the arrival into Santiago with Jorge and Carlota. I would regret that I had allowed my ego to dictate the finish of the Camino,"

"Yep and from what I understand, you are family and there is nothing greater than that." James added, "You are an amazingly strong-willed woman, like none I have ever known; don't let ego get in the way of making your decision. You do not have to prove anything to anyone, including yourself. Who knows, maybe you will write a book one day on the honesty of the Camino and the choices you are forced to make. It would be a bestseller!"

"Alright then, no more on the fence, I have it sorted. Having started this journey more alone than I have ever been and now to be surrounded by love is what I have been walking forward to. I am love and I can be loved. To walk into Santiago with that love is the ultimate destination. It signifies my growth and eagerness to spend the rest of my life mastering my own connection."

"That is why I love you Tess, though I know we can never be. You have given me the gift of understanding a connection I have never known. Our short time together before you left, I knew how special you were. To have been part of your Camino in words has been the best thing I have experienced. I have grown to love the incredible, thought-provoking, beautiful woman that you are and you have made me a better man. I have no regret, only mountains of gratitude."

"James you a dear man. Our paths crossed for a reason and we both know why. We will always have a connection, besides who the hell else will buy this supposed book I'm going to write?" Saying goodbye to James as Carlota and Jorge walked into the room, I felt happy having finally reached complete peace in my decision. James was a knowledgeable man with wonderful insight into what I already knew. He had understood my need to walk the whole way, but also stirred in me, would it be worth losing my family's embrace in the end. The answer was simple, No.

"Good morning sleepy pilgrims." I smiled as they sat down. "I just spoke to James and he sends his best."

"That is great," Jorge said, still trying to wake up. "He is our Saint James you know."

"Holy shit, I never thought of that." I laughed as we acknowledged the different names of the Camino; The Way of Saint James, The Way, and, of course, our personal favorite, The Other Way.

"Jorge you are pretty smart for a brother you know."

"I try little sister, I try," he replied.

I hoisted up the purple monster, which would now just slide into place and, with a see you later, I was off. With my choice securely made to walk into Santiago with them, I needed some solo time prior to the beginning of the end of my Camino. I knew they would be doing the same, with the unspoken words between us.

The sun had come out yet again so I stripped off my light coat. I passed a cement marker with the appreciated yellow arrow and carved beneath was 120km to Santiago. It was a long way from the 790km sign in Roncesvalles after climbing the Pyrenees. I smiled. Walking down what I was sure was a cow trail, I contemplated that the early arrival into Santiago would mean four days before my flight home—so, Muxia?

My thoughts were quickly interrupted as a large herd of cattle came ambling toward me, herded along by a beautiful old woman in her house apron. Not sure what to do, I just kept walking as the cows parted to let me through until I came head to head with a massive bull. His nose ring glistened in the sun as he licked his lips, both of us stopped within three feet of each other. He was enormous, taller than me, with shoulders displaying his ranking of the herd. The animal looked me in the eye and I did the same, there was no fear, only the acknowledgment of sharing a path. A cow mooed in protest of the hold-up and broke the spell between us and he politely stepped to my right side and walked on. That was an experience I was likely to never have again.

The path snaked along a mountain meadow overlooking a large valley below. Beyond the next visible mountain range would be Santiago. I listened to my music and my favorite song started to play, Stairway to Heaven by Led Zeppelin. Standing alone on a rock, overlooking the beauty, I sang along with Robert Plant at the top of my lungs. "There's a feeling I get when I look to the west and my spirit is crying for leaving. In my thoughts, I have seen rings of smoke through the trees and the forest will echo with laughter." How I loved that moment that was mine and mine alone. The lyrics felt as if they were written for me. The release of passion was euphoric, I thought of how everyone should sing at the top of their lungs once in a while. Looking ahead, off to the side, in a separate meadow, I saw the couple I had watched over breakfast that morning. They lay tightly together against their packs talking in their secret language of love. It gave me such pleasure to see them again. I took a moment to make a heart on the trail with stones and flowers for Carlota and Jorge, should they see it when they would walk by, sharing the love. I found a clearing off the way, so I sat in the dry autumn leaves to admire the view of the valley. I was just far enough off the path that when Eva and Alex walked by they did not see me. Instead, I saw them, my friend and her husband, holding hands and he was laughing at whatever she had just said. Love was everywhere in full force for me to see that day.

After a short rest, I gathered myself up and resumed walking. Reminded of all the up climbing I had done the day before, I was now on the descent. All along the Camino, I remembered little notes of insight pilgrims would leave for one another in every spot imaginable. There was one area weeks ago, where an entire section was littered with handwritten notes secured on the ground with stones. My favorite had been, "If not now than when." I felt then, coming down the mountain, I had the insight to give to others. I grabbed the black marker I had bought to decorate my gypsy stick, and scribed on a perfectly flat rock, "To find the love you must put complete trust in the journey, for it will find its way to you." Satisfied I stood up and there were Carlota and Jorge standing there watching me.

"Well, that worked," I said. Both looked down and read what I had written.
"All love, right little sister?" asked Jorge.
"Yes, Jorge, every love that is possible."

As we headed off together again, the Camino began to change as predicted. Now walking amongst us were pilgrims in bright, clean clothes with small day packs. Groups in matching shirts, to represent families or clubs, passed at alarming paces.

"Is this the start?" I asked .

"It is I'm afraid. Some start farther back then the 100km mark. "We still have another smaller mountain ahead and most of them will bus that section. We still have some of our quiet time," replied Carlota.

"I'm grateful for that," I said. A continual stream of the weekend walkers marched past us in all their gaiety and joy. It was a long weekend in Spain because All Saints Day was approaching and traffic was busier than I'd seen on the whole Camino. It was exactly how it was described to me in Ponferrada, except I would not see the magnitude of it because of the choice I had made. Rather than resenting these pilgrims, I found I could appreciate their excitement at the start of their own Camino. They were fresh, rested and super happy as they walked past us with the constant Buen Caminos to us. I remembered when I started a month ago and cyclists passed me and I had thought they had it so easy. After having conquered my fair share of kilometers on Nelson, I had an admiration for those that rode the whole way. Those weekenders passing me now may not have had six weeks carved out of their lives to walk, so my only thoughts were good for them for doing what they were doing.

We hunkered down for the night in a lackluster town—I couldn't remember the name—in a cold, concrete, expensive, modern hostel. It was so foreign to me and I was feeling a disconnection to all the authentic experiences I had been grown accustomed to. Jorge and Carlota were feeling it as well and had retired to their room early for the night. I was hungry and went into the town center and found a restaurant with outdoor seating. I wondered how I was ever going to sit inside to eat when I got back home to winter. Ordering a meal and a small bottle of wine, I pulled out my phone for nothing else better to do in this weird place. The screen displayed four missed calls, the caller unknown again just like in Ponferrada. What the hell! Halfway through my meal, the phone rang again displaying the mystery caller, I hesitated for a moment and answered my phone for the first time in over a month.

"Hello…?"

"I am looking for Tess Corps," the official-sounding Canadian voice came over the line.

"This is she," I replied.

"My name is Constable Allan Brookes, with the Langley RCMP. Are you the owner of a '96 VW Cabriolet, license plate GPE624, red in color?" he asked. I was immediately worried, I had left my car in the care of Sarah and my mother's panic button was pressed.

"Oh my god! Is my daughter alright?"

"I believe so, we have found your car abandoned in a ravine in the Fort Langley area. It showed evidence of being stolen; we have tried to reach you for several days," he said.

I briefly relayed to the constable my current situation and passed on the telephone numbers of Sarah as well as my older son, Mike. Constable Brookes assured me he would contact them and have the car returned. After the call, I messaged both kids to ensure they were safe and with news of the car. Sarah was quick to respond that neither one of them was in danger and that she was very sorry. She had parked the car at her work to store until I returned and had not even noticed it was missing. I told her not to worry, I was just happy they were okay. There was not much else I could do sitting halfway across the world, so I let it go.

Finishing my dinner, I thought about the days leading me slowly back to the world that was not the Camino. I was dealing with each awakening reality one at a time, regardless if I wanted to or not. Walking back to my stark, perfect hostel, I knew that my other life was approaching and I felt a sense of sadness that the Camino was going to end soon. Was I ready? Did I have it solid in my heart and soul what I had come for?

My God, the third stage transition was engulfing me.

NOT EVERYTHING IS A SIGN

It was the 30th day of the month and my 33rd day on the Camino, we were three pilgrims walking together—lots of threes. I had always been interested in numbers, especially reoccurring ones, and the number three had always been one that followed me when things were on a good turn, so waking up with those thoughts deemed a good day ahead.

Not long after we started the day I drifted ahead, or my family drifted behind, it wasn't exactly a conscious thought, we just traveled at a distance when we needed to. Climbing what was to be the last mountain, I fell into a comfortable easy rhythm. My feet knew their job, my strong legs pulled me forward and my shoulders leaned into the hill making the climb easier. Cresting the top, I took not a rest but a moment to sit and admire the beauty. I thought of the many paths I had walked, not only physically but also mentally and spiritually. This time in my life had been exactly what I needed to be forced to dig deep and to discover the wonderment of my very being. We lived in a world of people, including myself, that all asked the same questions, "Am I okay?" or "Was that okay?" I had my answer and I was and would be okay. Knowing that came clear to me on that last mountain and I was ready to go home and embrace the journey of my life that was still yet to come. I had trust and belief in myself and had put expectations aside to be open to whatever would be.

As I stood up, I made eye contact with the ever-so-quiet professor. It had been a while since our paths had crossed.

He slowed his already ambling walk and spoke to me, "Good you are going to make it."

I smiled at that peaceful man. "I certainly am."

"I worried about you; do I know you have a strength that I admire.

185

Enjoy the last days, you have earned your right to be here." He waved his arm in the direction of all our surroundings. It was then he smiled at me, for the first time, and walked away. I felt touched by the man of very few words and the impact of his selected thoughts meant so much. Sometimes we exist and think no one is watching, but they are. He had taught me humanity.

Several hours later, I came across a grove of unique trees with naked trunks that felt like silk as I caressed their smoothness. One particular tree was directly on the path and, asking permission from the tree, I inscribed with my marker for all the pilgrims to see, "Trust the Journey." That would be my final thoughts I could share with those who would pass.

"Another good one little sister," Jorge said, as the two arrived at where I stood. It no longer surprised us to find each other in a random spot. "Of your many gifts to us Tess, this I know will always be the most remembered," said Carlota.

Together we left the tree behind and began our final mountain descent. As Carlota had predicted, the weekend pilgrims had not climbed the mountain, rather bussing around it, which allowed us the solitude we all deeply desired. On the heavily wooded forest trail, I was acutely aware that since walking the Galicia region, I saw no cafés or rest stops as there had been before and none of us had snacks or even water. We had roughly walked 17 kilometers and there was still another 8 kilometers before the outskirts of Sarria. The only option was to put our trust in the way and as always it would provide what we needed. Out there in the woods, in the middle of nowhere, behind an ancient stone wall was a rustic pilgrim retreat of the sort. We smiled knowingly at each other before we entered under the arched passageway. Someone was playing the guitar and there was a table laid out with fruits, cookies and water, welcoming us into the tranquil space. Tom, an Australian, came over and introduced himself to us and asked if we would like a Café con Leche, officially, I was transported to heaven.

Sipping on my homebrewed coffee, I took in the setting of the retreat. Under the skeleton of an old barn, there was a corner spot set up to meditate, old couches to rest, stacks of books, paper for to write prayers and messages to be left, and stones to write words of wisdom. Jorge and Carlota had gone off and were engaging in Spanish with another pilgrim, while I sat on a rug playing with an orange mama cat and her three cuddly matching kittens. A young man came over and asked to look at my gypsy stick resting against my pack. In the time since I had found it, I had been adding with my marker simple drawings and phrases all over as well as tying mementos to the top. The young man carefully picked up my stick gently turning it around many times to look at all its creativity

"This is a good stick," he said looking with intent. "I make many sticks for those who come here and need strength to finish the Camino. I put a great deal of energy into my work for them."

Placing my stick back against my pack, he said, "Your stick is very charged; I can feel your strength through it." I was already aware of how special it was, but his words took it to a whole new level.

"Thank you."

Carlota came over and joined me on the rug.

"Tess, you will like the story of what I have just learned. The property is owned by a couple from Galicia, they believe in giving back to the Camino so they made this retreat for passing pilgrims. It is completely run by pilgrims who have walked in to rest and end up staying for periods of time." Carlota pointed to Tom. "He came eight months ago has no plans to leave yet. His new friend, Ben, arrived one week ago and is also staying for a while. They commit to harvesting the land to keep it self-sufficient so that this beautiful place can provide. There is no money exchanged here, only kindness," said Carlota.

"This is the first time in Spain I could easily stay behind," I said. "I feel I could be part of giving back to the Camino and sharing the valuable lessons I have learned. It is a sanctuary of peace, freedom, and creativity." Carlota smiled in understanding. The time came for us to leave and I quietly thanked the Way for providing exactly what we had needed.

Reluctantly, we returned to the forest with another six or so kilometers to Sarria. The final descent of the last mountain was almost complete. Somehow, I got separated from my family after a very large herd of unmanned sheep made their way past at a full gallop. I was alone, crossing over a small road, when I saw a sign indicating 600 meters to the left was the House of Magic. Of course, I was intrigued but thought better not to leave the way, not sure why, so I just kept walking. Further along, there was a fork in the footpath in front of me, with another sign indicating the House of Magic was only 315 meters, encouraging further in small print only a six minute walk. With two invitations I couldn't resist. I made the turn off the Camino thinking, what's another few more meters at this point? I was faced with an uphill climb and far more than six minutes, of course, to arrive at the house with slight trepidation. More of a feeling than what I was seeing, it was just strange. Stopping at the end of the small driveway, I questioned whether I should go in. I saw three backpacks against a stone fence and felt better about exploring. A small haphazard entrance with empowering rocks of all kinds displayed on a table drew me in. Once inside, the very dark, converted barn, I could hear voices in the background, but my eyes were drawn to the powerful artwork hanging on the stone walls in the narrow passageway. I looked closely at each piece with no particular subject but of colors in every hue dancing on the canvas. With a soft light source shining upon each one, I was mesmerized. Walking the length of the wall admiring each piece, I found an open doorway and peaked in a room completely dedicated to meditation. The drapery covered the hard walls and ceiling and the stone floor was not visible because every inch was covered with pillows and cushions of all sizes. Having satisfied my curiosity about the House of Magic, a different kind of magic I supposed, I turned to leave, only to be approached by a man inviting me to join them for tea. I politely declined explaining in broken Spanish as best I could that I had only come in to explore. Walking back to where I had come, I had liked the artwork but I laughed out loud at myself, a new lesson had been learned, not everything is a sign.

I returned back to the Camino and in the distance ahead I could just briefly make out Jorge and Carlota rounding a bend. Quickening my pace a bit, I thought I might catch up to them. When I arrived at the same bend they were nowhere to be seen. I stepped up my pace even faster but to no avail, I was not going to find them. With reluctance, I slowed down and remembered the Camino gives you what you need and for me it was further solitude. My feet also reminded me the pace was too much at that point in the day and I resumed the comfortable stride Jorge had taught me weeks ago.

The final approach into Sarria came with not the usual end of the day feeling of joy but a bittersweet reminder that I had just finished the last day of unspoiled nature. The next day would be the start of "the other way" and would bring me closer to the end of the Camino.

I was hot and tired and I stopped at the first café on the busy street, thinking somehow I must've passed Carlota and Jorge and at any moment I would see them arriving. I enjoyed a café con leche, oh how I would miss those, and a freshly baked neapolitan, as I swatted away flies that insisted on attaching to my dusty, sweaty body. Next to me sat a group of pretty women around my age from California. I could not help but listen to their conversation; more because it had been so long since I had heard my native tongue spoke so fluently. They were discussing their excitement of starting the walk to Santiago the following day and receiving the Compostela. I admired them kindly as to how fresh they looked in makeup and stylish walking clothes and their combined exhilaration of the weekend. We as individual people find our joys the way we need to, that is every right given to us, I was thinking. I finished my treats and, with no signs of my troop, I swung my pack on and continued on following the arrows to old Sarria, not before wishing the ladies a sincere Buen Camino.

My feet had swelled up in the time I had sat in the café, which in turn felt like walking with weights attached. The distance was longer than anticipated, not yet seeing any familiarity with the old part of town. I walked amongst concrete apartment buildings and businesses, crossing several busy streets and alleys. Eventually, the arrows brought me to a beautiful wide moving river and a pedestrian bridge to cross to the other side, where many lovely cafés and restaurants bordered the river and the atmosphere was festive and lively, I would definitely come back later for dinner there. As Carlota had been finding me inexpensive hostels each day, I knew it would be somewhere in the old section of town. Unfortunately, the herd of sheep separated us before I could get the information of where I would be sleeping that night. Still following the arrows, I kept walking until they directed me to climb a huge stone stairway.

"REALLY!?" I said out loud. Starting the climb up, my steps were steady and slow, my pack pulling me backward with every step on my pudgy feet, but I eventually reached the top. There, I found the old part of town and, with relief, I plunked down on the last step to rest. On my left was a small shop with the Wi-Fi sticker on the window so I pulled out my phone to see if Carlota had sent me a message regarding my lodging. Ping! There was an email from Jorge from hours before. I sure wished I had checked earlier because, according to my GPS, my hostel was at the bottom of the stairs.

"Bloody Hell! Too funny," I laughed at myself.

There had been a lot of hilarious moments on the journey and that moment was definitely one of them. I stood up to go back down the stairs I just laboriously climbed. If my feet and legs could talk they would have certainly said, are you kidding?

I rang the door of the hostel and a motherly woman opened it and welcomed me with a hug. Up the elevator, we went and I was let into more of an apartment than a customary hostel. It was cozy and well-lived in. Down a small hole, counting only three rooms, I was given the key to a delightful, small room with a beautiful view of the river. I thanked the woman and sat on the bed, sliding my feet out of the sandals and removing my pack from my back with relief. There was a soft knock at my door and I opened it and the woman handed me a hairdryer with a huge smile. That object in my hand was so foreign I had to laugh. It was something I certainly took for granted back home and a symbol of my daily world coming back. Taking full liberty of the communal bathroom, as it appeared no one else was staying in the apartment, I had a long, cleansing shower and blow-dried my hair for the first time in weeks.

I ate a dinner of calamari and cheese alone by the river later, happily people watching: couples walking hand-in-hand or sharing romance over a glass of wine, teenage girls gossiping and giggling over shared desserts, parents chasing after toddlers and men playing chess at tables set up just for that. It was all picturesque and I felt joy to be in the hub of it all. Something inside me changed that night, the sadness of leaving the Camino was gone replaced by happiness and peace.

THE OTHER WAY

I woke up the following morning after a glorious peaceful sleep and nothing hurt, whoopee! The excitement of arriving in Santiago was mounting. The other way would commence later in the afternoon when Carlota, Jorge, and I would board the bus to bring us 20 kilometers outside the city and from there we would walk to our destination. A decision that was both difficult and easy for me.

I spent the morning strolling through old Sarria, after climbing the huge stairs again. The volume of pilgrims had increased substantially, mostly consisting of fresh weekend warriors committed to the 100 kilometer mark. There was an obvious distinction between the two types in our pack sizes and faces. The long-term pilgrims, like myself, had weather-beaten faces, tanned, dry skin, and cracked lips. The look in their eyes showed more to the story of the time we had spent on the Camino, with unknowingly demure. The others were crisp in appearance and full of fresh energy, ready to embrace their new adventure.

Approaching the end of the old section, I found the arrow that would have directed me to Santiago and out of respect for those that would travel it and also honor to myself, I walked it for three kilometers. On the return back, I was intrigued by the ruins of a fortress tower off the road with no signage. It was surrounded by a stone wall that in no way was permissible to enter, but I found a crumbling section and was quick to scamper up over. I was never able to resist an opportunity to explore a mystery, especially one such as that. Once inside the five foot walls, visible by no one except five grazing sheep that were surprised to see me, I took it all in. The small courtyard was surrounded by gnarly old trees and had been taken over by grass and weeds, with the relic of what was left of the small building.

191

Upon further exploration, I found on the backside of the building, stone steps leading to the top of the tower and, without a second thought, I made my way up. There I discovered square columns I could only imagine were built for guards to hide behind. I stepped up to one and saw the view, it was more than I could have expected. I could see far across the Galicia mountains that I had climbed and beyond all the kilometers I had walked. It seemed so fitting to be right where I was. Standing in the sun that filtered between the columns, I embraced that moment and all that I had accomplished to be there. I reflected on how different the two separate cultures of Spain had been. The first half I had seen massive cathedrals adorned in gold and glass, private castles, expensive vineyards, and experienced the well-known history of that part of Spain. For the most part, it had an aura of wealth in various degrees. After Astorga, it had all begun to change slowly. I had passed through simple villages of days gone by but yet the history was in the present and very much alive. The churches, like the people, took on simplicity with only wooden crosses adorning the walls behind the altars. No more of the random cafés I had come to love, they had become far and few in between. Instead I had experienced mountain climbs with vistas as far as the eye could see and enchanting magical forests. The comparisons of the two cultures in the country I had come to love had been unexpected. But it was the Galicia region that would hold a place in my heart, it would be a connection to a way of life I could understand, simple.

I had not yet connected with my family that day and I was feeling the awareness of our time together coming to an end. Going down the formidable staircase again, I went back to my room to retrieve my pack and to check out. I walked along the river, exploring. I stopped for coffee and to check my email. Carlota had sent many messages wondering where the hell I was and that they would be at a certain place for lunch at 1 PM—I had 15 minutes to find them. Putting the name of the café they were in my GPS, grateful for the app, it directed me once again to do the large stone mountain of stairs for the fifth time in 24 hours, which I did laughingly.

"So you thought you lost me I bet?" I said jokingly when I found the two of them happy and rested.

"You know we try but you are like an octopus stuck to our face!" Jorge teased standing up to embrace me and one of his brotherly hugs.

We quickly caught up on our adventures from the day before, with Jorge raising his concerned eyebrows at my recap of the house of magic. I learned that while I was trying to catch up with them at Mach 10, I had actually walked right past the café they were sitting in. "We kept asking people if anyone had seen the little Canadian writing in a book," Jorge smiled.

"Haha that's funny," I replied.

A boisterous older Australian woman, whom we had briefly encountered at a rest stop a few days prior, took the table next to us. We had all admired her for her age and walking the pilgrimage, though she was a tad annoying in conversation. Her voice was far louder than needed, and she spoke her opinions, that did not sit well with us, fast. We smiled, acknowledging her, but did not say a word as she talked. Jorge had the same look on his face that he had had with the one-tooth Portuguese man, to remind us, girls, do not engage. It was an inside joke between us three as we stifled our laughter. She spoke regardless of our silence,

"You know, this pilgrimage, the Camino thing is nothing more than a long, bloody walk. I'll be glad when it's over."

My eyes bugged out of my head as I kept them focused on Jorge who gently kicked me under the table. With no response from any of us, the woman stood up with a humph and walked away.

"Holy shit," I breathed again, "She is one helluva character."

"And she is doing the Camino, her way," Carlota laughed.

"Good thing we kept our mouths closed, it was a conversation I could do without," replied Jorge.

"I have learned that I don't have to give myself to someone I don't want to. The choice is mine and not everyone is going to jive. I have always tried hard to connect with everyone in my life and when it failed I have always thought it was me. I now know different and it makes it okay." I said.

"There are enough people in this world that will give and receive us just the way we are. We too have learned this freeing lesson," said Carlota. We all nodded our heads knowing how lucky we had been in meeting each other.

At 2 PM it was time to commence the "other way" for the three of us. Descending the staircase for the last time, we walked through downtown Sarria towards the bus station. Somehow on the Camino, dressed in my attire with my pack, I looked normal but at the bus station, I just looked homeless. An observation that made me giggle, thinking of the phrase, don't judge a book by its cover. Tickets in hand, we climbed aboard into the comfort of an air-conditioned, modern bus for the two and half hour ride. Carlota and Jorge curled up together and promptly fell asleep with their hands intertwined. The two of them had become one without for a second losing who they were as individuals. I turned my head to the window, resting against the glass watching the unaccustomed speed of passing kilometers. Old guilty feelings tried to encroach my thoughts but I was quick to suppress them. Dozing off, I dreamed of the moment I would walk into Santiago.

"Wake up little sister, we are here," Jorge said as he gently roused me. Climbing off the bus into the dark, it drove away leaving us in a cloud of dust.

"There is a man coming to get us," organized Carlota announced. "Any time now." After twenty minutes had passed I said, "Right, like almost flat?" That sent all three of us into fits of laughter. When the headlights came out of the dark and pulled up beside us we were still laughing.

It was Halloween Eve and the place Carlota had arranged was extraordinary. Pulling up to the hostel in the dimness of the night, all we could see was a home built in classic Gothic architecture.

"Did you pick this place on purpose Carlota? Being Halloween?" I asked her.

"No not at all," she replied with the same look I had of WOW.

Once inside, the decor complimented the outside with Gothic medieval touches everywhere. We were escorted to the common area and sat on a high back throne-like chairs upholstered in rich red velvet as we were served wine, cheese, and various meats. The host, Arne, joined us and spoke slowly in his broken English.

"It has been some time since I've had pilgrims for the night and I welcome you to my home. Not many stay the night this close to Santiago, it is such a pleasure."

Carlota briefly relayed our plan of the "other way" and he responded with a knowing smile and full of support. He told us a history of his home, as we indulged in his offerings, that it had been passed through the family over many centuries.

"It has always been a refuge for pilgrims and will always be. Originally it was one of the few places to rest spiritually before a pilgrim completed the journey. With the modernization of the last part of the Camino, the way had been slightly changed and now my home is not as easy to find, but you three did," he said.

I was marveling at the story and thinking how if I had not chosen to take the other way, I would not have been part of the historical place.

We were showed to our rooms for the night, which also fit the Gothic theme of the place. A huge dark wood canopy bed with a crushed blood-red coverlet awaited my tired body, not before I had a bath in the gold inlaid clawfoot tub in the corner of my room. I loved my last night on the Camino, in that place on Halloween eve, it could not have been more perfect.

EARLY START

Slowly waking up from my deep sleep to the sound of chimes from my alarm, I realized today was the day. Thirty-five days ago I had left St-Jean-Pied, France, full of anticipation, adventure and excitement, completely unaware of the road that lay before me. Now a mere 20 kilometers away, I would reach the destination of St. James's tomb and I pay my respects as millions of pilgrims had done over many centuries. I reached into my pack beside my bed to read the words I'd carried since the beginning, now fully understanding the author's message"

"For untold thousands of years, we traveled on rough paths, not simply as peddlers or commuters or tourists, but as men and women for whom the path and road stood for some intense experience: freedom, new human relationships, a new awareness of the landscape. The road offered a journey into the unknown, leaving behind expectations, that could end up allowing us to discover who we were." - John Brinkerhoff Jackson.

I had found who I was with the feeling of wholeness and completion. The seeker I had been had become the finder. The emptiness I had felt for years was now full of hope, belief, and renewed passion.

I assembled my pack in the familiar rhythmic way, aware that this would be my last day on the Camino. My emotions felt muted, I just wanted to be in that moment with no expectations. The last day started, like so many others with a tired body, sore feet and hefting my pack on my back, but it would be so very different. Coming out of my room, I went directly outside to wait for Carlota and Jorge. We were starting out earlier than ever because we wanted to arrive for the 1 PM Mass and the swinging of the Botafumerio. There would be no coffee or breakfast until later, so I sat in

196

the darkness drawing on my smoke, putting into perspective some simple thoughts.

I remembered, when I watched the movie eight months ago, feeling the pull. Sitting with Him, who, as it turned out, I hardly knew at all, as he expressed that He could see me doing that—walking across Spain. What was first a seed, then the loss of a relationship, loss of a job, and above all loss of hope, delivered an opportunity. Then by some miracle, the plan started. I declared to the universe my intent, scraped together money, maxed out the visa because, "if not now, then when?" It could not wait. I left my family, friends and the life I knew behind to go on the epic adventure of a lifetime with a tattoo inscribed on my forearm, "Trust the journey." It was my ticket to self-discovery. There I sat, six weeks later with 780 km behind me, almost to the destination of the adventure.

I had walked the Pyrenees mountains of France, across northern Spain, over rivers and streams, through and above magnificent valleys, over many mountains, seeing and believing in it all. I had experienced the kindness of strangers and moments of instant connection with fellow pilgrims. Even though I walked alone most days, I never was. For I was surrounded by the beauty of nature and the Buen Caminos along the way. Some days were insanely tough but those days always followed by gifts of all sorts, that would inspire me to keep going. The sun had burned my skin, sweat drenched my clothes, and the kilometers of blisters had destroyed my feet, yet my perseverance had conquered it all, and it was almost over.

"Are you ready little sister?" Jorge gently touched my shoulder to bring my thoughts to him.
"I am Jorge… I really am."

We started our walk silently in the morning darkness through a misty eucalyptus forest, which gently dropped condensation on our faces. The sun would rise soon to welcome us to our glorious day. It would take around four hours to reach Santiago, and I relished in my last quiet steps to the forest. Carlota and Jorge walked slightly ahead, holding hands as they were symbolically completing their union as a couple. All the previous meadows, forests, horses, cows, sunny days, conversations, mystic moment, religious revelations, glasses of wine, outdoor cafés, languages, blisters, stones, bandages, boots, uphills, downhills, sacrifices, and effort of the Camino would soon be over. In return, I knew I would feel the personal accomplishments and growth and it would have been worth it.

SANTIAGO

In just under two hours, we arrived at Monte do Gozo, translated as the Hill of Joy, and I knew why immediately—far off in the distance, I could see the three spires of the Santiago Cathedral. We stood in silence, side-by-side, each of us in a personal thought or, in my case, looking for the first time at what represented everything about the way of St. James and the Camino.

"It's really there," I whispered.

The emotions I had been carefully guarding were coursing through me, frantic to come out. No, not yet, I want to feel this feeling as long as possible.

"This is a good day to finish the Camino," Carlota said, giving my hand to squeeze. Renée, my friend who had walked the Camino earlier that year, had told me there was a spring nearby, which traditionally pilgrims would go to wash in preparation for their arrival into Santiago. I had thought previously I too would follow the tradition, but standing on that hill in the soft breeze, I had never felt so cleansed and fresh in my life.

"We should get moving," Carlota suggested, "We want to make the pilgrims mass for 1 PM".
"Yes for sure," Jorge said.
"Come on let's go, we can't sit around all day, it is off to Santiago we must go." I laughed.

As we came down from the hill of joy, I walked in front of Jorge and Carlota on the groomed asphalt walkway. Passing a metal light post, I saw

198

the Celtic symbol of strength and encouragement drawn in black marker. The three connected spirals centered by a triangle was the final gift to me from the Camino. I took a photo and smiled secretly, I would have this tattooed next to my trust the journey, to signify the three of us. It represented not only our love for one another but the strength and courage we had overcome together. I said nothing to Carlota and Jorge of receiving the gift, knowing one day I would be able to show it to them.

My final steps of nature ended with me on the sidewalk of a busy street, directly across from a large metal sign, SANTIAGO DE COMPOSTELA. I waited for my family and together arm in arm we crossed over. Touching the sign, I could no longer hold my emotions in and tears of so many feelings rolled down my cheeks. Both Carlota and Jorge came, with tears in their eyes also, and we embraced openly at the busy intersection. We had made it.

Jorge took both our hands and turned towards the city. "Come on let's go find a café con leche and have a moment before we go to the Cathedral."

We sat inside a small café, as a light rain started to fall and we shared her feelings of happiness to have arrived and discussed what it will be like to not walk every day.

"Some mornings I would wake up and my legs would be vibrating with the need to walk," I said.

"I would feel the same, especially on the days we rested," Carlota added.

"For me," said Jorge, "I wanted the forest, to get into the forest. I will always remember the trees and the gift I received from them." Carlota and I smiled at Jorge, jokingly teasing him he is now one with the trees, though, in sincerity, we both knew it was true.

The rain stopped and the sun was doing its very best to shine through the dissipating clouds. Walking out of the café, the smell of wet concrete swirled into my nostrils. Following the now very official arrows, and scallop shells in the sidewalk, we found our way to the entrance of old Santiago. Stepping into the ancient world, the streets narrowed, long-standing stone buildings welcomed us, and there were pilgrims in every direction. Every person we passed offered a congratulatory smile of our arrival. There was power on the cobblestone I walked that relieved all the pain and replaced it with love for my courageous body and cleared my soul. The Cathedral spires were within our vision with every corner and the anticipation in our eyes glistened in the sun that shone with greatness. The streets with no geometrical pattern snaked toward the Cathedral.

We rounded the last corner and 200 feet in front of us was the historical archway that would take us into the Cathedral Square. The simplistic ancient stone entrance to the end of the Camino reminded me of the much smaller archway I had walked under in St. Jean Pied 800 km ago. I could hardly believe I had arrived. Carlota, Jorge, and I linked arms and walked beneath it as a dream became a reality.

There it stood, everything I had visualized, read and talked about: St. James Cathedral of Santiago in all its historical glory. We stood in the square and before it, tears erupted without shame, the celebration was ours in that amazing moment of life. As much as I wanted to do this trip alone, having those two amazing souls beside me to share that moment was perfect. We had become, in our time together, a family in every description imaginable.

UNIFICATION

Standing before the grand staircase up to the cathedral, I looked high above at the peak and saw the stone figure of St. James. I had completed, The way of St. James, The Camino, and, as fondly called amongst pilgrims, The Way. The bells of the church started to ring, echoing a beautiful sound to welcome us to the mass that would begin shortly.

"Let's go, Tess, it is time to experience the high mass and you will now see the Botafumeiro," said Carlota, excitedly taking my arm.

Passing through the grand entrance of the wood double doors that stretched to the sky, I could feel the presence of all the pilgrims that had walked this stone before me. Once inside, I could not have been prepared for the interior grandness of the cathedral of Santiago. The detailed painted ceilings high above me, the massive round columns throughout supporting the many archways, the ancient tapestries of stories hanging from the stone walls, my eyes could not focus fast enough. Then I saw the elaborate altar, the most elegant of any I had ever seen in my lifetime. It stretched not only high but deep into the church, ornated with statues of gold and marble and at the back and, in spectacular wonder, was St. James. The Cathedral was shaped like a cross with three sections filled with pews that met in the middle before the altar. Above that hanging, from gold-colored ropes, was the Bumafumeiro, even bigger than I could have imagined. I couldn't wait to see it swing through the cathedral, the incense burning inside cleansing the pilgrims below.

Carlota came to my side. "Come, Tess, let us go sit down with Jorge, it will begin soon." Taking my eyes off the Bumafumeiro I followed her to

201

the pew where Jorge sat waiting with a smile. Sitting down, Carlota whispered to me that from her last visit she knew the best place to sit to watch the mass and the Bumafumeiro swing, I nodded in gratitude. Reaching into my pack for the pilgrim prayer—by Fraydino from Santa Maria la Real—which I had received in O Cebreio from Carlota, I read it again:

Although I may have traveled all the roads,
Cross mountains and valleys from East to West,
If I have not discovered the freedom to be by myself,
I have arrived nowhere.

Although I may have shared all my possessions
With people of other languages and cultures
Made friends with pilgrims of a thousand paths,
Or shared albergues with saints and princes,
If I am not capable of forgiving my neighbor tomorrow,
I have arrived nowhere.

Although I may have carried my pack from beginning to end
And waited for every pilgrim in need of encouragement,
Or given my bed to one who arrived later than I,
Given my bottle of water in exchange for nothing;
If upon returning to my home and work,
I am not able to create a brotherhood
Or to make happiness, peace, and unity,
I have arrived nowhere.

Although I may have seen all the monuments
And contemplated the best sunsets;
Although I may have learned a greeting in every language
Or tasted clean water from every fountain;
if I have not discovered who is the author
Of so much free beauty and so much peace,
I have arrived nowhere.

If from today I do not continue walking the path,
Searching and living according to what I have learned;
If from today I do not see in every person, friend or foe
A companion on the Camino;
If from today I cannot recognize Spirit/God
As the one spirit of my life,
I have arrived nowhere.

The word of God was unified with Spirit in my heart, no longer an argument for two separate entities, just simply a feeling to connect to all.

The mass started with a haunting yet serene piece of music played by a layman in the procession, it was amplified due to the design of the Cathedral. Carlota and I looked at each other with eyes of indescribable emotion as silent tears freely flowed down our cheeks. The parade of many priests and laymen past by us, and in the center, carried on a platform was the bust of St. James. The music heightened and I looked forward to the altar, the swinging of the Bumafumerio had begun. My heart was pumping against my chest for the epic moment I so needed to experience. The eight Tirabuleiros and red-colored robed men each took a rope and began the dance that engaged the Bumafumeiro to begin the greatest pendulum movement I had ever witnessed. As the ark grew to over 200 feet above my head the music reached a crescendo. The moment I was in would remain ingrained in my memory forever.

The music slowed as did the Bumafumeiro and the service proceeded. The priests sang as they continued the walk towards the altar, each taking a seat in their respective positions of the church. One priest stood up to the pulpit that towered over the congregation and, in beautiful Spanish, began the service. I felt the words he said and understood through those feelings, what was being spoken. Clasping my hands together, I prayed with all my soul to those that had stood the test of time with me and to those that had done me wrong. I prayed for my three children, that their lives would be full of as much love and joy as possible. And then I prayed for the man whom I cared about for so long, that He'd find His way to the same piece I now had. As a baptized Catholic, I took communion though it had been many years since I had done so. But on that day, I accepted God as the Universe and the Universe as God and it was a sacred ritual for me that I now could fully understand. When receiving the communion, the priest look so adoringly into my eyes, I felt his unconditional love in my heart and soul.

THE COMPESTELA

With the service completed, we sat a few moments longer before Jorge announced, "Okay girls, no more tears, it is time for celebration! *Vino blanco* and lunch it is."

We enjoyed a festive feast and many toasts: to our backpacks, the mountains, her feet, showers, clean clothes, the people we had met, our walking sticks, and above all to each other for having found one another. It was a wonderful time for all three of us, though knowing our time together would soon come to an end weighed on our hearts.

Exiting the restaurant after lunch, we stood outside not sure what to do next, there was no more required walking and we felt a bit lost, then Carlota's eyes twinkled.

"Let us all go to the pilgrim office and get our last stamp from Santiago on our Camino passport."
"Yes, it is too bad about the Compostela but you will have a stamp, Tess, woohoo," teased Jorge.
"Oh just shut up big brother," I joked in return. " I have long gotten over the Compostela thing, see look I am a pilgrim dancing in the street."

With that, I bounced around with the purple monster on my back dancing to the music only I could hear. Jorge and Carlota just stood there laughing while onlookers smiled at the energy surrounding us.

"Come on silly woman, let's go get that stamp," said Jorge.

We arrived at the pilgrim office to find the long lineup we anticipated was nowhere to be seen. Showing our passports, both country and Camino, to the sentry at the door ,we were directed through an entrance to a long counter with volunteers. I walked forward, all serious in that place of honor, to a delightful happy young man.

"Hola, hello," I said.

He replied in perfect English, "Welcome and congratulations on your completion. May I see your Camino passport please?"

I handed my credentials over to him smiling with pride as he proceeded to leaf to the many pages of stamps inside.

"Where did you start from?" He asked

"St-John-Pied."

"What was your pilgrimage, religious or spiritual?"

I thought for only a moment and responded with the truth, "Both."

At that moment it really hit me of all the lessons I had learned and I wept silently. The young man stood immediately and reached across the counter to give me a hug.

"You have accomplished an amazing journey."

Returning to his seat, he reached under the counter and proceeded to fill out a beautifully-adorned, ivory-colored document using a calligraphy pen. Finishing his work he handed the paper to me,

"You are acknowledged for all that you came for."

In my hand was my Compostela. I whispered my gratitude to the thoughtful man as I walked away with that final gift from the Camino. By letting go of expectations, I had lived the experience of every moment and it was far more rewarding, the proof was in my hand.

Outside I found Carlota and Jorge also with the Compostelas they had received.

"Can you believe this?" I said waving mine in the air.

Carlota was grinning. "I know we all felt it was not important now that we have it I think it is a great unexpected gift. Am I right Jorge?" she asked.

"Yes my beautiful woman, it is the certificate of our Camino union," he said, looking adoringly at her. "And Tess, yours is to authenticate that you are compassion, love, and strength."

"Thank you, Jorge, your words mean a great deal."

The unexpected had truly brought the Camino to its peaceful end.

We meandered back to Cathedral Square, finding it full of newly arrived backpacks of every color attached to people from all over the world. The air was festive and full of life and energy as we stood amongst it, enjoying that moment.

Knowingly we turn to each other, I said, "It is time…"

I looked into the faces of the two people I felt I had known for a lifetime. First I reached out to hug my Camino brother and whispered in his ear, "Always love her deeply, she is by far the greatest gift of your life."

"You dear Tess, of all people, know I will," he replied hugging me harder. "Take care of yourself, my little sister."

I turned to Carlota, both of us tearing up as we embraced each other fiercely. "We will see each other again," I said.

"Yes and another journey," said Carlota.

"Thank you for all you did for me," I said through tears and laughter.

"And I thank you for sharing yourself and your wisdom with us."

Jorge gently tugged Carlota's arm and began the slow step that would physically separate us. I watched them as they walked away hand-in-hand, pausing briefly before the archway we had entered together not that long ago to look back at me. And then they were gone.

THE CALL

After my Camino family's departure, I sat down right where I was in the middle of the square. I stared up at the beautiful Cathedral and felt overjoyed by everything. Beside me lay my gypsy stick, heavily worn at the bottom with an eroded section from where my hand had gripped it. I had doodled so many pictures and thoughts all around it and the treasure had fought to the top of so many memories, it would be a souvenir I would take home. Underneath it was my Compostela resting on the sacred ground. I lay back against the purple monster and watched the faces of the pilgrims arriving, smiles so big, laughter and tears of relief and joy radiated everywhere. Groups were dancing, some threw their walking sticks in the air, while others stood in peace. An older man I had seen a few weeks back arrived, dropped his pack, and did an airborne somersault!

Several hours went by before I stood and heaved on my pack and set out to find my lodgings for the night. Dropping off the purple monster in my room, I set off to explore the ancient city. Life was everywhere, music on every corner, cafés overflowing with pilgrims in celebration, and history everywhere I looked. I came across a lone pilgrim looking lost and I approached to welcome him. He was not sure where the square was and I happily pointed him in the right direction and, with a small smile of relief, he headed to the end of his Camino. I stopped at an outdoor tapas bar where I ordered food and wine and set about messaging my children and James to let them know I had arrived at the end of the Camino. I posted on Facebook to let my friends know and the responses came flowing in immediately. First my children, full of pride and excitement that I would be home soon. The likes and comments flooded my post, filling me with more love. It was my own perfect celebration at the little tapas bar with my large glass of *vino blanco*.

The bells from the Cathedral rang as I walked in the night air and found myself at the back entrance of the church, where a small door was wide open. I ventured in. I had wanted to see the tomb of St. James earlier but the lineup had been long but now there was no one there. Going down the stairs into the catacomb, I respectfully knelt on the small bench in front of the tomb to honor the saint that had inspired the pilgrimage I had completed. I returned back outside into the night and breathed in deep, feeling more alive than I ever had.

I could hear the ocean calling me and I lifted my face to the sky with a serene smile, I knew I would have to heed the call.

EPILOGUE

Three days later, in Muxia, I sat in front of an alluring turbulent ocean with waves crashing with great force at my feet. This was the end but also a promise of a new beginning. Having lived a lifetime in six weeks, I knew there would be still be so many more journeys ahead but I had confirmed, within myself, that I was strong, courageous, and full of love. That the many paths ahead, no matter how difficult, I would tackle and let go. For every time the Way appeared to be difficult, it would always provide a reward on the other side.

In our day-to-day lives, we have a tendency to procrastinate the big mountains that challenge us and to wallow in grief that the mountains created. We fabricate what is coming, rather than experience the moment we are in and often experience disappointment. The Camino was always there and it was impossible to put anything off, nor speculate what was around each corner. Regardless of the blisters, exhaustion or disappointments, the day still had to be completed. We live our lives looking for instant gratification to our problems, but we have failed to acknowledge the wine grapes to be eaten, the spring-fed fountain to drink from, or the patch of grass overlooking a valley to rest at. So focused on the end result, we miss the gifts along the way that support and guide us through those times.

I was taught this lesson every day on the Camino. Life will give you struggles, where often there is no end in sight. However, when you look back on past experiences, the end always arrived. By letting go of something you cannot control it brings you freedom. But if we fail to believe in this, the disparity is the only thing we have left. Despite the steep mountains, long, desolate stretches, and sharp descents over rolling rocks, I always got there. In this chaotic demanding world we live in, time for self is of the utmost importance. To discover your own self-love, you simply have to go there, either physically, mindfully, or spiritually because eventually, it all unites you to who you are. The key is to look for yourself and not involve any outside opinions or influences. Look at the world only through your eyes and see the beauty that will evoke your spirit and keep you mindful.

The heart and the soul will reward you with an abundance of peace. Yes, you will be vulnerable, but in doing so, the answers you are seeking will find you.

Start by going for a walk.

ABOUT THE AUTHOR

Tess Corps writes and lives in a cottage in a meadow in central British Columbia, Canada. Her passion for nature, life and being the very best she can be is what makes her tick. Discovering the extraordinary in the ordinary fuels her adventures, whether across the world or riding her Harley Davidson through the mountains near home. She is a mom of three outstanding adult kids and best friends with a remarkable artist. Living a life of simplicity, gratitude she knows it will always be okay.

Tess's story will not end here.

Made in the USA
Middletown, DE
23 December 2021

56934027R00130